PRODUCTION
of
ISOTOPES

A portion of the Proceedings of the All-Union Scientific and Technical Conference on the Application of Radioactive Isotopes·Moscow, 1957

IN ENGLISH TRANSLATION

Springer Science+Business Media, LLC 1959

ISBN 978-1-4899-4721-5   ISBN 978-1-4899-4719-2 (eBook)
DOI 10.1007/978-1-4899-4719-2

$ 50. 00

Library of Congress Catalog Card Number: 59-14487
Copyright 1959 by Springer Science+Business Media New York
Originally published by Consultants Bureau, Inc. in 1959.
Softcover reprint of the hardcover 1st edition 1959

PRODUCTION of ISOTOPES

CONTENTS

CONTENTS (continued)

THE DEVELOPMENT OF ISOTOPE PRODUCTION IN THE USSR

V. V. Bochkarev, E. E. Kulish and Iu. S. Frolov

Artificial radioactive isotopes were first prepared about 25 years ago, but their exceptional value as a powerful tool for investigation and their diversity and usefulness in practical applications are difficult to evaluate at the present time. The interaction of the radiations from radioactive isotopes with material such as living tissue, their capacity for penetrating a material and the possibility of using isotopes as tracers offer very wide possibilities for their use in science and technology.

After a short period of research and development in new methods of preparing isotopes, these methods were mastered at a very rapid rate.

At first isotopes were used for solving a series of research problems and were prepared in small quantities with laboratory neutron sources and accelerators. It was only with the appearance of nuclear reactors and powerful cyclotrons that conditions became appropriate for the preparation of sufficiently large amounts of isotopes for their massive application in the national economy to be possible.

The requirements of scientific research and medical institutions for isotopes increased at such a rate that it became necessary to organize a systematic supply of isotopes and, consequently, their regular production. Among the first isotopes to be supplied regularly were the radioactive ones, I^{131}, P^{32}, Fe^{59}, C^{14} and Co^{60}, which had already been used in industry for a long time.

At first the isotopes were prepared by laboratory methods. Thus, P^{32} was prepared by irradiating an iron phosphide target with deuterons in a cyclotron and extracting the product chemically. I^{131} was extracted in small portions from solutions containing uranium fission products. C^{14} was isolated in microcurie amounts as CO_2 from potassium nitrate solutions, irradiated in a nuclear reactor in sealed glass ampules. Na^{24} was prepared by irradiating a target of sodium chloride crystals in a cyclotron.

In principle, many methods of preparing radioactive isotopes are possible. The great majority of them may be prepared either in a nuclear reactor (neutron irradiation and extraction from fission products) or in a cyclotron.

Although cyclotrons are more generally applicable for preparing different isotopes, nuclear reactors have the very great advantage of being more productive and the techniques for preparing isotopes by neutron irradiation are simpler.

For these reasons, neutron irradiation in a reactor was used primarily in organizing the mass production of isotopes. The extraction of isotopes from fission products and targets irradiated in a cyclotron is employed as an additional method for preparing those isotopes which cannot be obtained in a reactor at all or with the required specific activity. A general diagram of radioactive isotope production is given below. As the diagram shows, three main stages are described: a) preparation of the raw material; production of isotope—intermediates; c) processing of isotope—intermediates into compounds and objects containing radioactive isotopes.

Preparation of Isotopes in Nuclear Reactors

The preparation of radioactive isotopes by neutron irradiation in a nuclear reactor is based on nuclear reactions by so-called "pile" neutrons, whose spectrum contains mainly thermal neutrons with an energy of the order of 0.025 ev,and also fast fission neutrons which have not been moderated.

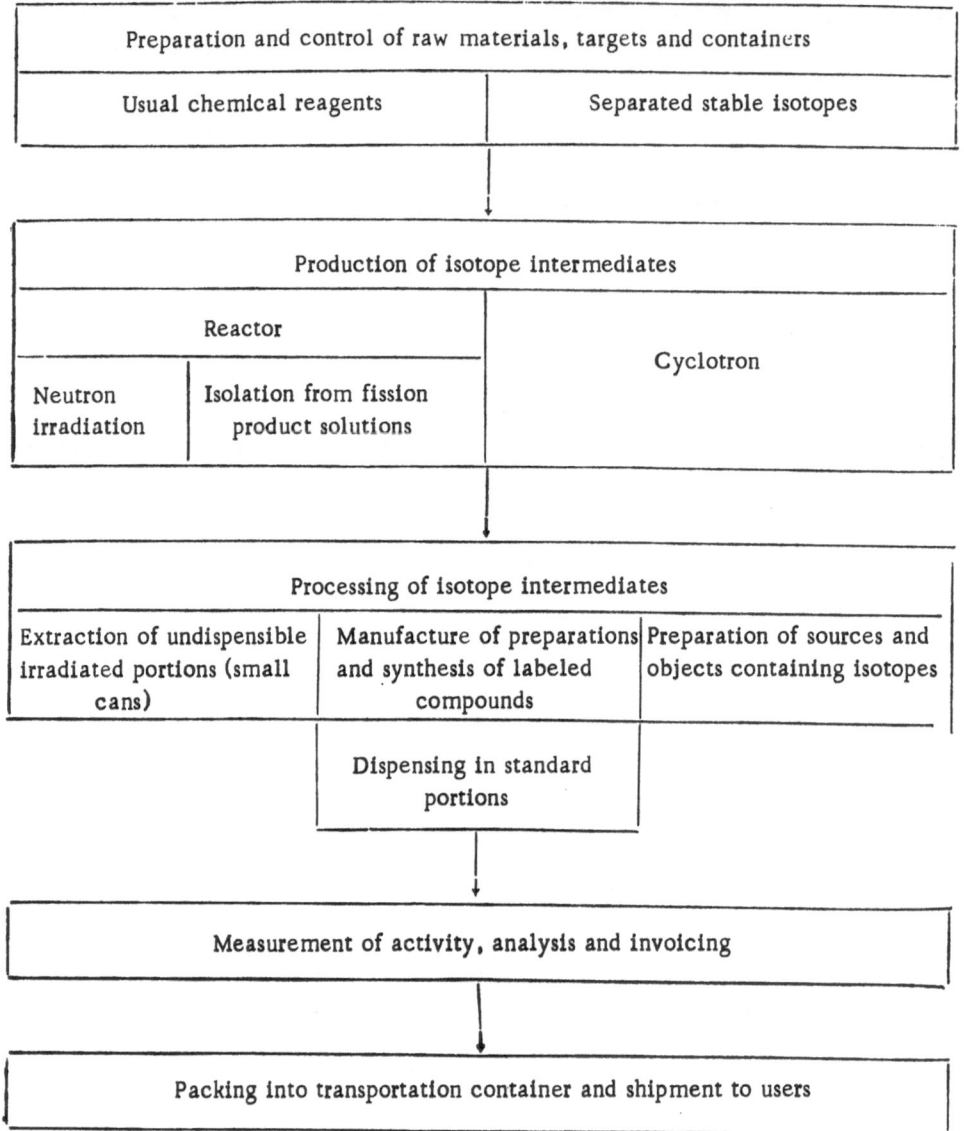

The simplest and most commonly used type of reaction consists of neutron capture by a nucleus followed by emission of a γ-quantum, the so-called (n, γ) reaction. This gives an isotope of the same element but with a mass number one unit greater than that of the original isotope.

This reaction gives, in particular, such isotopes as Na^{24}, Co^{60}, Ir^{192}, Au^{198}, etc. Generally speaking, radioactive isotopes of all elements except helium may be obtained by this reaction; however, some of them have either too short a half-life or are formed in such minute quantities that it is impractical to prepare them by this method.

The second reaction is the (n, p) reaction in which a nucleus captures a neutron and emits a proton. This forms an isotope of another element with an atomic number one unit less. This reaction can be carried out with thermal neutrons in only a few cases in which the energy of reaction is positive, and it gives such important isotopes as C^{14} and S^{35}.

Besides the (n, p) reaction, the (n, α) reaction is also used for preparing radioactive hydrogen (H^{3}) from lithium.

Quite a few reactions can be carried out practically with fast neutrons as in certain reactors the fraction of fast neutrons reaches 30-40%; for example, irradiation of sulfur can give P^{32} with a high specific activity.

The isolation of isotopes formed in a decay chain, for example:

$$Te^{130} (n, \gamma) Te^{131} \xrightarrow{\beta} I^{131}$$

$$Sm^{154} (n, \gamma) Sm^{155} \xrightarrow{\beta} Eu^{155}.$$

is also of great importance in production.

Other methods are also used.

The preparation of radioactive isotopes is now undoubtedly one of the important uses of nuclear reactors.

In the USSR various types of reactors may be used for the production of isotopes and, as a result, isotopes may be obtained with the minimal expenditure. At the same time it is possible to use high-density neutron fluxes for the irradiation. Usually we use fluxes of 10^{12}-10^{13} neutrons/$cm^2 \cdot$ sec to obtain isotopes, but in individual cases fluxes of higher density are available.

Although different reactors with certain constructional peculiarities are used for isotope production, one can speak of certain general elements of isotope production by neutron irradiation.

Raw Material

The selection of the starting chemical reagents and their preparation are of extreme importance in the problem of isotope production.

The material must satisfy many serious and occasionally contradictory requirements. Two of them are unconditional: the material must be safe under irradiation and of maximum purity so that impurity activity formed will not mask the main activity of the isotope.

Special technical conditions worked out for the reagents used during isotope production require a high purity of the raw material. Thus, for example, metallic antimony should contain less than $10^{-4}\%$ of cobalt and less than $10^{-3}\%$ of iron; ferric oxide should contain less than $10^{-5}\%$ of cobalt, thulium oxide should contain less than $1.5 \cdot 10^{-3}\%$ of holmium and less than $3 \cdot 10^{-3}\%$ of lutetium, etc. Satisfaction of these requirements demanded considerable effort in developing methods of purifying and analyzing the raw material.

Isotope	Normal raw material			Enriched raw material		
	Specific activity, mC/g	Impurity activity		Specific activity, mC/g	Impurity activity	
		element	%		element	%
Fe^{59} ..	5	Fe^{55}	220	460	Fe^{55}	0,7
Sn^{123}	27	Sn^{113}⎫		320	Sn^{113}⎫	
		Sn^{117}			Sn^{117}	
		Sn^{119}⎬	49		Sn^{119}⎬	1,7
		Sn^{121}			Sn^{121}	
		Sn^{125}⎭			Sn^{125}⎭	

The use of separated stable isotopes is also of great importance in the production of radioactive isotopes. Besides considerably increasing the specific activity, the use of an enriched raw material makes it possible to obtain a preparation which is isotopically purer and this is illustrated by the examples given in the table on the preparation of radioactive Fe^{59} and Sn^{123}.

In addition to Fe^{59} and Sn^{123}, it is possible to produce a large number of other isotopes at the present, using enriched raw materials, e.g., Fe^{55}, Cd^{115}, Ca^{45}, etc.

Containers for Irradiation

At the beginning, when the volumes of the loads were extremely small, the chemicals were irradiated in cylindrical graphite containers which housed the sealed glass ampules with the material to be irradiated. From 1949 onward we have used aluminum slug-containers of different construction, which has made it possible to use the facilities for loading and unloading uranium rods in the reactor, and this has simplified the work of the personnel.

The use of such containers, hollow cylinders with screw tops, made it possible to increase the output by improving the utilization of reactor channels and the volume of the containers themselves. However, the fact that the containers were not hermetically sealed gave rise to a series of difficulties. Then containers which were hermetically sealed by argon-arc welding were used and this made it possible to decrease the wall thickness and increase the useful volume and, consequently, increase the amount of material irradiated at the same time. As a result of these improvements, the relative useful volume of the containers was tripled in comparison with the first ones.

Changes in the container construction naturally necessitated a change in the methods of loading the raw material. Thus, if the raw material is intended for chemical treatment and the preparation of labeled compounds or objects, it is loaded in large portions either directly into the container in bulk or in briquets, or into large aluminum cans; if, however, the sample is to be delivered to the customer without any preliminary processing, then it is loaded into the slug in several small aluminum cans. This simplifies the work in the second stage of isotope production as it partly eliminates dispensing of the radioactive material and decreases the loss of the active product. Modern handling techniques are used in operating with slug-containers.

Irradiation Time

During the first years of organizing production, when 10-15 isotopes were produced and these mainly had relatively long half-lives, the differentiation between irradiation times relative to half-life was of no great importance. Later, when the number of isotopes produced increased considerably, in order to increase the throughput of the reactor channels and to raise the total output of isotopes, efficient irradiation times were established. At the present time, four main groups of isotopes may be listed in order of irradiation time.

The first group with an irradiation time of 15-24 hours, includes isotopes with a half-life of several hours to three days, for example, Si^{31}, Na^{24}, Cu^{64}, Au^{198}, etc.

The second group with an irradiation time of 30 days, includes isotopes with short half-lives (from 3 to 30 days), regardless of cross section (P^{32}, Cr^{51}, etc.).

The third group with an irradiation time of three months, consists of isotopes with a half-life of over 30 days (Fe^{59}, S^{35}, Ca^{45}, Hf^{181}, etc.).

The fourth group with an irradiation time of over six months, consists of isotopes with very great half-lives (Co^{60}, C^{14}, Cl^{36}, etc.).

One should note that this is, of course, a conditional separation and serves only as a general indication. There are cases where, due to specific properties of the material or isotope, it is prepared under conditions which do not correspond to the given classification. Thus, for example, Ta^{182} and Ir^{192}, which should belong to the third group according to their half-lives, are classed with elements of the second group due to the large activation cross section of the starting isotopes (Ta^{181} and Ir^{191}).

The improvements in reactors during the years of isotope production development have made it possible to increase the isotope output severalfold.

Thus, in the preparation of C^{14}, the change from irradiation of calcium nitrate solutions in small glass ampules to the use of long aluminum, hermetically sealed containers filled with compressed calcium nitrate and special separating equipment has made possible an increase in the activity of C^{14} per gram of barium carbonate from several microcuries in the first years to tens of millicuries at the present time, and a sharp increase in the total production. Its growth during recent years is given below.

	1953	1956	1957 (planned)
$BaC^{14}O_3$ production, curies	2.2	11.2	50.0

It should also be noted that the cost of this form of production has been decreased by a factor of approximately 30 in comparison with the original cost.

Other Methods of Preparing Isotopes

As has been noted previously, a considerable number of very valuable isotopes may be isolated from fission products. The most important of them, from the point of view of their prospective use, are Cs^{137}, with a half-life of 33 years and a γ-radiation energy of 0.67 Mev, and Sr^{90} with a half-life of 20 years and a β-radiation energy of 2.18 Mev.* Of the other fission products, Pm^{147}, Tc^{99}, Eu^{155}, Ru^{106}, etc., are of great interest.

The technological methods being developed for the preparation of these isotopes are based on the effective use of waste from atomic production — waste industrial solutions. Procedures have been developed for the separation of 14 different isotopes from the mixture of fission products. Fresh solutions are used for the isolation of short-lived isotopes, and fission product solutions that have been kept for 2.5-3 years are used for preparing long-lived isotopes.

In developing the technological separation scheme, special attention was paid to the need for large amounts of radiochemically pure Cs^{137} of high specific activity. The method developed ensures the production of radiation sources from cesium with a radiochemical purity of 99.9% and a specific activity of 60-70 curies/cc.

The other long-lived isotopes, Sr^{90}, Ce^{144}, Pm^{147}, Eu^{155}, were separated by a complex technique, which consisted of the successive separation of the above isotopes by first concentrating the activity and then isolating the isotopes by various methods.

The fission isotopes being supplied at the present time have no carrier, a radiochemical purity of 98-99%, and not more than 1 mg/mcurie of salt impurities.

The second direction for the effective use of the isotopes found in atomic waste is as important. We are considering now the supply of unseparated concentrates of fission isotopes for industrial use as powerful radiation sources. Previously such concentrates were used in the form of solutions, but at present techniques have been developed for preparing concentrates of unseparated isotopes in the form of solid sources with quite high specific activity that are stable during storage and use. The application of such sources is definitely very promising.

The available cyclotrons are used for the preparation of certain isotopes such as Na^{22}, Mn^{52} and Mn^{54}, As^{74}, V^{48}, etc. The supply of these isotopes is still limited.

Processing of Isotope—Intermediates

Let us examine the next stage in the production of isotopes before they are used.

When the containers with the materials have been irradiated in a reactor or the targets in a cyclotron, they are transported to storage facilities and are kept there for a more or less long period of time, in some cases until the short-lived components of the sample and the activated casing have decayed, and in others until the time when they are required for further operations.

Storage of Containers with Isotopes and Work with Sources.

The storage of hundreds of highly active containers with isotopes and thousands of separate sources has become simple and safe with the construction of special water storage-reservoirs. Even highly active γ-sources present no danger to the personnel under a thick layer of water (from 1.5-3 m), while the simplicity and convenience of working with them with special equipment and good visibility made it possible to cut down the cost and raise the output of the corresponding products.

The reservoirs are constructed in such a way that besides the storage of containers in individually labeled compartments, the containers can also be opened (cut), sorted, and the activity of the extracted samples

---

*Radiation from the daughter Y^{90}.

measured and other operations performed under water. This not only reduces the use of bulky, expensive cells and boxes with remote handling equipment, optical systems and other complicated devices, but also protects the air of the room and the personnel from contamination with radioactive dust.

The containers with sources or samples, insoluble in water, are removed from the storage compartment to the section where they are opened by a simple cutting device and their contents extracted. The sources can either be stored under water in the appropriate compartment or be passed on, also under water, to the next section for activity measurement and invoicing.

The γ-radiation activity is measured by a relative method by comparison with standard sources. The apparatus used for this is an air-filled housing made from plexiglas or another light material, in the form of a cone,into whose base is set an ionization chamber; the standard source and the source to be measured are placed in turn at the apex of the cone on the outside, close against the thin wall. The readings are taken on an indicating apparatus above the reservoir.

After measurement, the sources are either placed in storage or transferred to the next section, where they are inserted into transport containers or special irradiation apparatuses, used in industry and in medicine.

The technique of charging the various apparatuses varies, depending on the purpose and the design, and we will not describe it in detail.

The containers with other isotopes, intended for dispensing or further processing, are transferred from the underwater storage by a transporting device into special "dry" chambers for opening and from there they proceed to further technological processing.

We should also note that not only the slug-containers and sources in ampules are kept in underwater storage, but also active solutions and powders — intermediates and finished preparations — though naturally not in water, but in cells, lowered into the reservoir along a tube in an air space situated under the water (shielding layer), where they are distributed in sections and rows with a special device.

Range of Radiation Sources and Its Growth

In the first years, only cobalt radiation sources in the form of cylinders of various diameters and lengths were issued. In subsequent years, they were issued as cylinders in aluminum cases measuring 5 × 5 and 10 × × 15 mm and later, also in the form of hollow cylinders 24 × 24 mm with a wall thickness of 4 mm, with a nominal activity of 0.1, 0.2, 0.5, 5.0, 20, and 250 g-equiv. of radium.

A range of cobalt needles and applicators have already been supplied for a long time for use in oncological institutes. Cobalt–nickel alloy wires of various lengths and activities (length from 5 to 60 mm and activity from 0.5 to 30 mg-equiv. of radium) are set, after neutron activation, in stainless steel cover-sheaths in the form of demountable (on a screw) needles and applicators. The sorting and activity measurements are performed in the water reservoir, but the setting of the active wires in sheaths and further processing are carried out on special apparatuses with shielding and semiautomation.

With a more precise definition of requirements and an increase in the demand, the range of cobalt and other radiation sources expanded and, to some extent, became standardized. At the present time, cobalt sources, apart from those of the oncological range, are issued in the form of wires with an activity of from 0.5 to 50 mg. equiv. of radium and as cylindrical sources and discs in metal sheaths with an activity of 0.1 to 400 g-equiv. of radium, of eleven different types. Sources of low activity will be issued in small sizes as cyclinders measuring 2 × 2 mm.

The data presented below give an idea of the increase in the use of radioactive cobalt sources over recent years.

	1954	1956	1957 (planned)
Total amount of sources manufactured from Co^{60}, g-equiv. of radium	16,760	64,000	300,000

In 1957 we will also supply sources in the form of Zn^{65} wire with an activity of from 0.5 to 4.5 mg-equiv. of radium and sources in metallic capsules with Cs^{134} – four types from 0.01 to 5 g-equiv. of radium – with Eu^{152} and Eu^{154} – three types from 5 to 50 g-equiv. of radium – with Tm^{170} – three types from 0.02 to 0.5 g-equiv. of radium and with Ir^{192} – six types from 0.01 to 20 g-equiv. of radium. The iridium sources will be made in standard sizes: 2×2 and 5×5 mm. Previously, Cs^{137} was supplied largely in the form of solutions and in only limited amounts as a powder; in 1957 it will be issued in ten types of cylindrical sources with Cs^{137} in a metallic capsule, with nominal activities of from 0.0001 to 20 g-equiv. of radium.

Some β-sources are also made from Tl^{204}, Sr^{90}, tritium and other isotopes, and also α-sources from Po^{210}. However, all the various requirements for sources are still not completely satisfied. In connection with this a list of the types of α-, β- and γ-sources has been prepared for general consideration to obtain the standardization required for planning and setting up automatic mass production.

Preparation Technology and Range of Preparations and Labeled Compounds.

Besides the issue of α-, β- and γ-radiation sources and the distribution of isotopes in the form of samples, irradiated in cans in standard portions, in recent years there has been a considerable growth in the production of various chemical compounds, inorganic and organic, labeled with radioactive isotopes, including a series of medical and other special preparations.

The processing of irradiated materials and the synthesis of labeled compounds is performed in special laboratories, to which the isotopes are transferred from the opening chamber. Here, widely different technological procedures are used for processing and synthesis.

In the first years we issued mainly the simplest radioactive preparations, generally metals and their oxides and salt solutions, frequently of insufficient specific activity and of low purity. As the purity of the raw materials increased, the processing and synthesis technology developed and the technological equipment of isotope laboratories improved, there was a considerable extension of the list and an increase in the quality of the preparations issued and it also became possible to produce them in considerably greater volumes.

In the first period, the simplest chemical procedures and normal chemical technology were used in the manufacture of radioactive preparations. When the issue of preparations was still low, there was a tendency to build general shielded cupboards with complex and mainly mechanical devices for remote control of the various processes and operations. Later in the production, we used to a considerable extent specially and appropriately equipped cupboards and boxes, designed for performing strictly defined operations with definite isotopes, with stationary, closed apparatus and equipment, which made it possible to carry out the different processes completely; cupboards with readily dismantled apparatus were also used. As far as possible, we introduced the direction of processes and operations by the use of pressure and vacuum, including the application of hydromanipulators, designed for precision work with small volumes, partly semiautomatic apparatuses, and also light, specialized tong manipulators and special remote devices (for opening and sealing ampules, etc.).

The methods and manufacture and preparation control technology were fundamentally changed so that a whole range of special methods was used – extraction, isotope exchange, electrochemical and chromatographic methods, etc. There was an extensive development in the synthesis of complex organic compounds, labeled with isotopes.

In view of the considerable growth in the range of preparations issued and the variety of the methods used, particular attention was paid to the efficient selection of raw materials. Thus, for example, for the manufacture of preparations containing P^{32} (with carrier), two forms of starting material are used – red phosphorus and phosphorus pentoxide. Although phosphorus pentoxide is less economical from the point of view of operating volumes in the reactor (the effective charge is only 40% of that attained in the irradiation of red phosphorus), it is considerably more convenient for the preparation of a large group of compounds. Thus, dissolving it in distilled water and subsequent boiling gives labeled orthophosphoric acid, which in turn can be readily converted into preparations widely used in medicine, such as disodium phosphate and other compounds. From irradiated red phosphorus we can prepare such compounds as phosphorus tri- and pentachlorides, sodium diethyldithiophosphate, the so-called preparation "Aeroflot," etc. P^{32} without carrier is prepared from irradiated sulfur. Now a total of 18 P^{32} compounds are issued and the total volume of these compounds required over the last five years has been 250-400 curies per year.

The choice of raw material and preparation method may be illustrated by the following example. Previously, preparations with Br^{82}, in particular $NaBr^{82}$, were obtained by neutron irradiation of bromobenzene, using the Szilard-Chalmers effect, i.e., in this case, the expulsion of active bromine atoms from the benzene ring with subsequent extraction of the bromine with alkaline solution. The yield of sodium bromide thus obtained was low, not exceeding 12-15%, and then it was only in alkaline solution. A higher yield may be obtained by the direct irradiation of sodium bromide, but then the sodium is also activated to form Na^{24} with a half-life of 15 hours, which makes it impossible to work with the preparation as labeled with only bromine (its half-life is 36 hours). Therefore, the preparation method was changed. Now we irradiate barium bromide (barium does not present practical difficulties in the work from radiation), which is then dissolved in distilled water and the solution passed through an ion-exchange column with a cationite in the sodium form (in other cases, the potassium or ammonium form, etc.); as a result we obtain a neutral solution of sodium bromide with an inactive cation. The yield here is more than 90% and the exchange process proceeds rapidly; the capacity is raised by a factor of six or seven and the quality of the product is better. In this way we can obtain any bromide labeled with Br^{82}.

One of the most important isotopes for practical use is C^{14}. As has already been indicated above, the starting material for the synthesis of preparations containing this isotope is barium carbonate. The preparation of carbon-labeled carbonates and bicarbonates of sodium and potassium from it was previously carried out by decomposing the barium carbonate with acid and passing the carbon dioxide liberated through sodium hydroxide solution. This gave a mixed product, which was always in alkaline medium. At the present time, the labeled carbon dioxide obtained is used in an isotope exchange reaction, e.g., $NaHCO_3 + C^{14}O_2 = NaHC^{14}O_3 + CO_2$, which proceeds in high yield without any isotope losses and gives a solution satisfying the given requirements.

Now we issue 88 different preparations of quite high specific activity, containing C^{14}, including many complex organic compounds. Since most of these syntheses are multistage and lengthy, whenever possible they are designed so that the radioactive isotope (this does not only refer to carbon) is introduced in the latter stages of the synthesis, to reduce to the minimum the number of stages with active product. On the other hand, to increase the possibilities of preparing different organic substances, we also prepare a series of compounds which are "key" ones for further synthesis. C^{14}-labeled acetylene is used for the preparation of a group of such compounds, in particular acetic acid, benzene and propargyl alcohol and their derivatives; we also issue labeled alcohols (amyl, isopropyl, etc.), acids (butyric, propionic, valeric, etc.), aldehydes (formaldehyde, acetaldehyde, benzaldehyde), insecto-fungicides (DDT), etc.

We also issue preparations such as benzocaine, aspirin, barbital, caffeine, etc.

High-purity sodium carbonate or bicarbonate in the form of compressed tablets without filler is used for the manufacture of Na^{24} preparations; this makes it possible to simplify the technological process (in particular, to exclude overfilling and weighing of the active product), to raise the purity of the preparations and, using special shielding equipment, to increase the capacity markedly.

We cannot consider other examples. We will only recall that in 1957, 284 forms of compound, labeled with radioactive isotopes, were issued. In particular, 42 different compounds labeled with S^{35} were issued, including sulfides, sulfates, xanthates, potassium thiocyanate and thiophos,*and also such preparations as methionine, vitamine B_1, sulfazole, sulfidine, sulfanilamide , thiourea, etc. Several compounds with I^{131} (without carrier) are issued, including diiodofluorescein (as a result of improved technology and equipment, the output of iodine preparations has been increased by a factor of 25) and preparations with Fe^{59} (including iron ascorbate), Ca^{45} and Cl^{36}; we also issue a series of compounds of short-lived isotopes (including a radiocolloid of Au^{198}), etc.

Stable Isotopes

In considering the development of isotope production, we should also examine in detail the improvements in the technology of enriched stable isotope preparation. However, the methods and designs of apparatus used in this work are so numerous that it is impossible to consider them in detail in our report.

There has been a parallel development of the physicochemical separation methods: diffusion and thermal diffusion, separation, normal and low-temperature fractional distillation, electrolysis, etc. Different designs of special apparatuses have been developed, making it possible to enrich in considerable amounts such stable isotopes as D, B^{10}, C^{13}, N^{15}, O^{18}, Ne^{22}, Kr^{86}, etc. The use of electromagnetic separating apparatuses made it

---

* Transliteration of Russian — Publisher's note.

possible to also obtain other enriched stable isotopes over a wide range of elements, including isotopes of the rare earths and platinum elements.

The growth in the consignments of enriched stable isotopes for scientific investigations is obvious from the data given below.

	1955	1956	1957 (planned)
Number of isotopes supplied	55	78	136
Isotope deliveries, g	2,500	64,100	90,000

A supply of various chemical compounds, labeled with stable isotopes, including D, B^{10} and N^{15}, has been organized.

Some Results and Prospects

The efforts of a large number of engineers and scientists in developing the technology and organization of isotope production over a number of years have naturally resulted in an increased output and improved supply of these products to scientific research organizations, medical institutes, and industrial concerns.

The output of isotopes in our country has increased continuously, as is shown by the following data:

	1954	1955	1956
Number of radioactive isotope consignments shipped, thousands	11.5	16.0	23.5

In the first quarter of 1957, 8,974 consignments were shipped to users. A considerable portion of these went to industrial concerns in the USSR who use isotopes systematically in their production methods, and to medical institutes, who use isotopes for diagnosis and therapy.

For several years, isotopes have been exported to China, Czechoslovakia, Poland, Hungary, Rumania, East Germany, Bulgaria, and other countries.

The increase in the isotope requirements and the need for preparing an ever greater variety of compounds with radioactive and stable isotopes and the mass production of a wide range of different types of radiation sources and articles with radioactive isotopes for special purposes (including medical) necessitate further extension of isotope production.

First of all, we should note that considerable attention is being paid to the extension of the range and increase in the output of compounds with radioactive and stable isotopes, required in various branches of science. In order to develop methods and techniques for synthesizing new compounds and preparations labeled with isotopes, scientific research organizations of the Academy of Sciences of the USSR and its affiliates, the Ministry of Higher Education and other ministries, are assisting in addition to laboratories of the Ministries of the Chemical Industry and Health. The projected range for 1958 envisages an extension of output to 360 compounds, labeled with radioactive isotopes, and the manufacture of 60 new compounds with stable isotopes; 137 compounds with C^{14} will be issued.

At the present time, an extension in the range and output of special radiation sources with radioactive isotopes is required for massive application in industry. In future years the demand for this production will undoubtedly grow, since on the basis of studies carried out on the possibilities of using isotopes in the national economy there will be a wide introduction of radioactive methods of control and direction of production processes; there will also be an extension of the use of radioactive radiation in medicine for diagnosis and therapy, and in industry for performing more technologically developed processes. At the present time, on the basis of future plans, developments are being made in the technology and automatic equipment for remote manufacture on a large scale of α-, β-, and γ-radiation sources with Co^{60}, Tl^{204}, Sr^{90}, Cs^{137}, Tn^{170}, Pm^{147}, Po^{210}, and other isotopes.

The technological basis of isotope production in the USSR is being extended on a wide front.

The increase in the number of concerns occupied with the output of isotopes makes it necessary to have a single organization dealing with the supply of products containing radioactive and stable isotopes for use in the national economy. This organization — a trust of the "Union-Reagent" of the Ministry of Chemical Industry of the USSR — should be the functional body, which makes it possible for a user to obtain any form of product with isotopes without directly approaching the concern which produces this product. We are envisaging the supply of isotopes not only according to a set plan, but also in response to direct orders during the year; these orders would be met in a short period from specially planned stocks of radioactive and stable isotopes. Together with this, the "Union-Reagent" trust would be responsible for informing users on available and projected products. The first step in this direction is the issue of the catalog, "Isotopes" in 1957.

Improvements in techniques and organization of production has made it possible to re-examine the cost of radioactive and stable isotopes. From April 1, 1957, there will be a considerable fall in the prices of such important isotopes as C^{14}, Sr^{90}, Cs^{137}, I^{131}, Au^{98}, etc. A separate payment for dispensing is also established, which makes it considerably cheaper to buy isotopes in large amounts. There is also a reduction in the price of powerful radiation sources of Co^{60}, Cs^{137}, Ir^{192}, etc., by which for some of them the cost per millicurie progressively decreases as the activity increases.

Further development work on the production of isotopes and the availability of a wide range of new labeled compounds, preparations, and articles with radioactive and stable isotopes and radiation sources provides our industry, agriculture, medicine, and also scientific workers with ever greater possibilities for the fruitful use of this form of atomic energy in the improvement of industrial methods and the development of science. The problem is to fulfill as rapidly and as fully as possible the directives issued by the 20th Congress of the Communist Party of the Soviet Union on the over-all development of the use of radioactive radiations in the national economy.

CERTAIN ASPECTS OF THE PRODUCTION OF RADIOACTIVE ISOTOPES IN A NUCLEAR REACTOR

E. E. Kulish

INTRODUCTION

The manufacture of products containing radioactive isotopes may be divided into two completely independent stages: a) the preparation of "isotope-intermediates" by neutron irradiation, i.e., of definite portions of the raw material, irradiated in a reactor; b) dispensing of "isotope-intermediates" and synthesis of various chemical compounds labeled with radioactive isotopes.

Below we will consider the main aspects of the technology of the first stage of radioactive isotope production— — the preparation of isotope-intermediates.

The second stage of the production and the preparation of radioactive isotopes from fission products are completely independent parts of the production with their own particular technological peculiarities and therefore we will not consider them.

The plan of successive operations in isotope-intermediate production is given in Fig. 1.

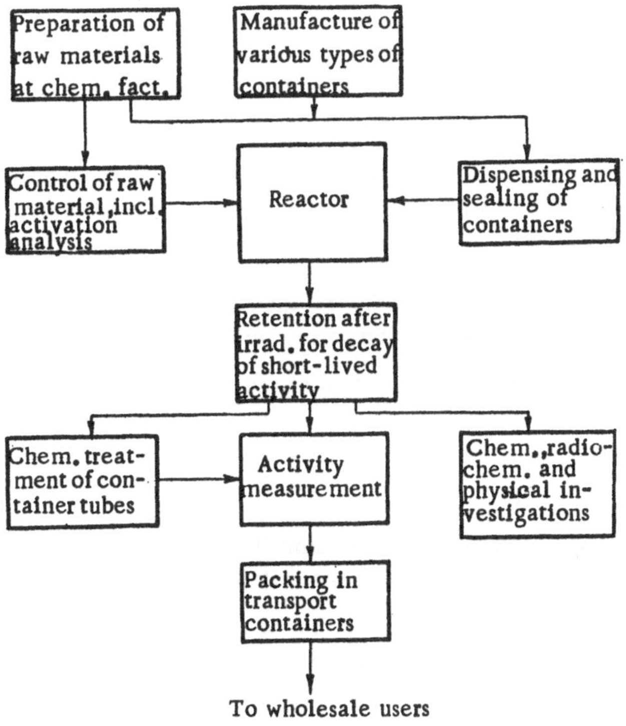

1. Raw Materials for Isotope—Intermediate Production and their Preparation for Irradiation

Various chemical compounds of different elements are used as raw materials for isotope-intermediate production.

In selecting a compound for irradiation, and also its degree of chemical purity, a considerable number of different factors have to be considered. Firstly, the raw material should contain the greatest possible amount of the particular element whose radioactive isotope is to be prepared. It is also essential that it should not contain elements with high activation cross sections, giving products with long half-lives, which would mask the activity of the main isotope, even when present in small amounts. The raw material must be stable to radiation and temperature, explosion-proof, readily convertible into other chemical compounds, etc.

During production, all the raw materials used are carefully checked for chemical composition and in some particular cases spectral, chemical, and activation analyses are carried out before irradiation.

Fig. 1. Plan of isotope-intermediate production.

11

Isotopes Obtained from Enriched Raw Materials

Radioactive isotope*	Raw materials irradiated	Isotope content, %	
		in natural mixture	in enriched raw material
Ca^{45}	Calcium oxide	2.06	80.0
Fe^{55}	Iron oxide	5.84	85.0
Fe^{59}	Iron oxide	0.31	70.0
Sr^{89}	Strontium carbonate	82.56	99.0
Sn^{113}	Tin oxide	0.95	40.0
Sn^{123}	Tin oxide	4.71	95.0
Cd^{115}	Cadmium oxide	28.86	95.0
I^{131}	Tellurium oxide	34.49	99.0
Te^{127}	Tellurium oxide	18.71	90.0

*Besides the most frequently used isotopes, which are given in the table, at users' requests, other radioactive isotopes, obtained by (n, γ) and (n, p) reactions in a nuclear reactor, may also be prepared from enriched raw materials.

To obtain preparations with high specific activity and increased radiochemical purity, raw materials are used which have been previously enriched in the corresponding stable isotope, before irradiation.

The table lists the main enriched isotopes from which radioactive preparations may be obtained.

Before irradiation, all the forms of raw material are packed into aluminum containers of various types. The type of container is determined by the purpose of the irradiated samples. Thus, raw materials for short-lived isotopes and samples for experimental work are packed in containers with readily unscrewed lids, since these containers are opened shortly after irradiation. In the case of long-lived isotopes, where the irradiated material can be stored for a long time without appreciable activity losses, the raw material to be irradiated is packed in hermetically sealed cans (Fig. 2), since they are stored under water in specially equipped reservoirs.

Each of these containers has a conventional marking, indicating the character and amount of material contained in it.

To facilitate further treatment of the isotope—intermediates, the raw material in each large container is packed in portions in separate small containers: aluminum cans and ampules, glass ampules, etc. (Fig. 3).

Raw materials, which are used for the preparation of a large number of intermediates in small amounts (e.g., raw material for C^{14}, Cl^{36}, etc.), are irradiated as compressed briquets, placed in special tubular containers.

Some forms of raw material, which have high activation cross sections, would be activated unevenly if packed tightly in a can due to strong shielding of the inner layers by the outer ones. Such substances are packed in containers with special holders to give the minimum shielding of some small portions by the others.

High-purity aluminum is chosen as the material for the cans and other containers as it does not form isotopes with a long half-life, is easily worked mechanically, and its corrosion stability is quite satisfactory, considering the comparitively short irradiation time.

Preparation of Beta- and Gamma-Sources

Radiation sources of certain standardized geometric measurements and activities are used for many special operations.

At the present time we are able to put out 46 types of source and a series of new forms are being developed.

In contrast to the standard portions of isotope—intermediates, raw material in powder form, for the preparation of sources, is extremely carefully packed in special aluminum ampules of various sizes, with tightly crimped-on lids.

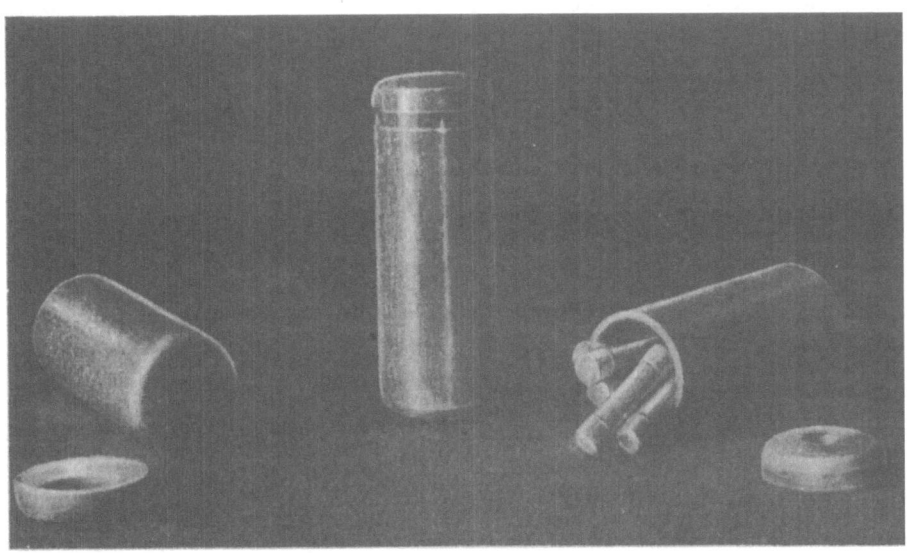

Fig. 2. Containers for raw material, intended for irradiation.

Fig. 3. Containers for small portions of raw materials.

In some cases the raw material is used in the form of plates and sections of wire of standard dimensions.

Before the aluminum can is welded up, all the sources in it are so arranged so that during activation the effect of one source on another will be minimal. Irradiations are performed for various periods, depending on the activity required and the nature of the raw material.

After irradiation, the cans with the prepared sources are opened and all the sources are checked for activity on an ionization chamber and then packed in transport containers for dispatch.

2. Irradiation Periods

Since the production of isotope—intermediates must make provision for further processing, the amount of isotope activity produced in one can is chosen for convenience of handling in subsequent stages of the production.

The specific activity, i.e., the activity of isotope per gram of original raw material, is set by practical necessity and production possibilities.

13

From the technological point of view, it is convenient to divide all the isotopes into four groups with respect to irradiation time:

1) Isotopes with half-lives of up to 3 days.

2) Isotopes, a) with small activation cross sections and small half-lives (but not less than 3 days), and b) isotopes with large activation cross sections and various half-lives.

3) Isotopes with low activation cross sections and long half-lives.

4) Isotopes with very long half-lives.

In accordance with this division, and also allowing for the definite requirements as regards specific activity, four irradiation periods are used: from 15 to 24 hours, and 1, 3, and 6 and more months.

After irradiation, the containers with the raw material are removed from the reactor with remote handling and either opened or placed in special storage spaces under water, which ensure adequate shielding of personnel from radiation.

3. Activity Measurements on Prepared Isotope—Intermediate Consignments

The activity of the irradiated raw material has to be determined before labeled preparations are manufactured in radiochemical laboratories.

Fig. 4. Transport containers.

Calculation of the activity from formulas is only accurate enough for evaluating the irradiation time for the various raw materials under the given conditions, but is unsuitable for determining the invoice data of isotope—intermediates, especially those which go directly to large-scale consumers (sources, standard portions, etc.).

In these cases, the characteristics of the isotope—intermediates have to be determined by direct measurement before shipment.

When the isotopes emit hard γ-radiation, the activity of the portions can be measured without extracting them from the containers, which considerably simplifies the measurements.

The activity is measured with an ionization chamber by comparing the activity of the containers with the γ-emitting raw material with the activity of a cobalt standard, previously standardized against a radium source. All the operations involved in the measurement of γ-emitters are performed under water.

The activities of C^{14}, Cl^{36}, and Ca^{45} are measured by the usual methods for determining β-active substances with thin-walled counters or 4π-counters.

The activities of other isotope–intermediates are checked by checking the integral neutron flux during the irradiation time by measuring the activity of indicator cans, containing standard portions of cobalt. Whether or not there are variations in the neutron density in the reactor and, if so, the character of these variations, are checked by measuring the activity distribution along a tube, in which are irradiated cans with raw material of extremely constant chemical composition.

After retention for 3-5 days for decay of the activity of the aluminum and short-lived impurities and for activity measurement, the cans with the isotope–intermediates are removed from the spaces where they are stored with special long handling instruments and placed in transport containers (Fig. 4), to shield personnel during transportation.

The transport containers were first made from lead, but were not durable or convenient and were replaced by combined containers made from a cast iron casing and a lead center with a cell to hold the irradiated can with the raw material. Transport containers weighing from 87 to 1200 kg are used, depending on the activity of the isotope–intermediate and the hardness of the radiation.

For the transportation of isotope–intermediates with soft β-radiation, such as C^{14}, Ca^{45}, etc., we use glass containers, packed in wooden boxes, which are filled with lead foil in some cases.

4. Preparation of C^{14} and Ca^{45}

Originally the isotope C^{14} was prepared in considerable amounts by absorption of the CO and CO_2 liberated during irradiation of very wet calcium nitrate in tubular containers in a nuclear reactor. Later it was established that with this system a large part of the active carbon was not liberated in the gas phase, remained in the raw material, and was only liberated after special chemical treatment.

In connection with this, a system was proposed using a filling of carefully dried calcium nitrate, which eliminated the liberation of carbon as gas and, at the same time, the complex system of gas take-off.

The operation plan for C^{14} and Ca^{45} production is presented in Fig. 5.

The irradiated tubular containers are transferred to an automatic cutting apparatus by remote control and are cut into pieces, which are tipped into a stainless steel tank. When filled, the tank is hermetically sealed and its contents treated with hot nitric acid solution.

The carbon dioxide containing C^{14} liberated during the decomposition of the calcium nitrate is trapped in a solution of sodium or barium hydroxide.

Experiments showed that when the process was carried out in this way, compounds of carbon, other than carbon dioxide (monoxide, methane, etc.), were not formed in appreciable amounts.

The apparatus for chemical treatment of the calcium nitrate was placed in a concrete shield to protect the personnel from radiation from the tank and the radioactive gases.

After removal of a sample for activity measurement, the filtered and dried barium or sodium carbonate is packed in a container and shipped to the users.

According to mass spectrometric analysis, the C^{14} content of samples prepared by this method reaches 20% and more.

The solution of calcium nitrate left after removal of the C^{14} is purified chemically to remove iron and aluminum and then the calcium precipitated from it as $CaCO_3$.

If necessary, the $CaCO_3$ is calcined to give calcium oxide.

The specific activity of regular $BaC^{14}O_3$ consignments is ~50 mC/g of $BaCO_3$ and that of $Ca^{45}CO_3$, not less than 15 mC/g of $CaCO_3$.

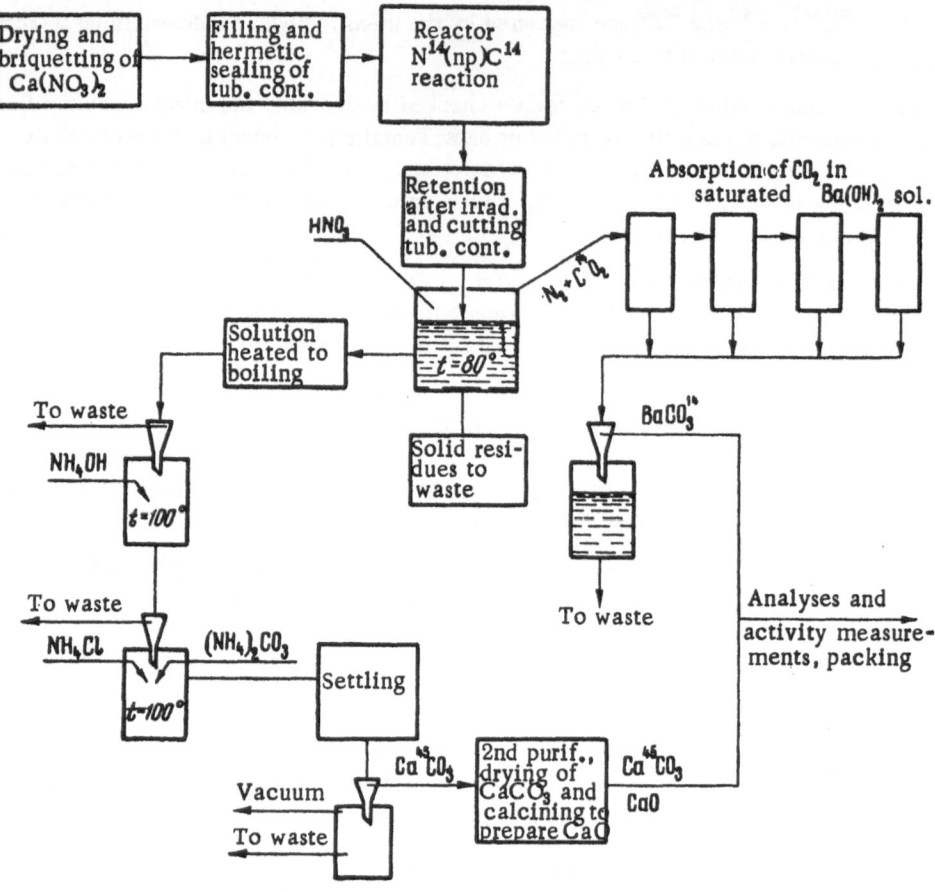

Fig. 5. Operation plan for C^{14} and Ca^{45} production.

5. Preparation of Cl^{36} and S^{35}

During the research and development period, the preparation of small amounts of Cl^{36} by the reaction Cl^{35} (n,γ) Cl^{36} consisted of the irradiation of separate portions of potassium chloride in standard containers and subsequent isolation, in general.

With an increase in the Cl^{36} requirements, it was more economical to irradiate the raw material in tubular containers and to isolate this isotope—intermediate in a special apparatus, as was done in the preparation of C^{14}.

Simultaneously with the isolation of Cl^{36} from the irradiated potassium chloride, S^{35} is also isolated and the preparation of the latter in large amounts is thus considerably simplified.

Great difficulties are caused by the strong corrosiveness of the solutions which imposes high requirements on the quality of the materials used in the construction of the separate sections of the apparatus.

SUMMARY

Above we gave a short account of the main aspects of the technology of isotope—intermediate production. A more detailed account is beyond the scope of our report and would hardly serve any useful purpose, as due to the great demand for isotopes and the extensive work with them in many laboratories, their production is undergoing considerable development and literally every day new improvements and changes in technological details are being introduced.

PRODUCTION OF RADIOACTIVE ISOTOPES IN A 10-Mev DEUTERON CYCLOTRON

P. P. Dmitriev, I. I. Zhivotovskii, N. N. Krasnov, I. P. Selinov and E.N. Khaprov

Certain radioactive isotopes which are of high value for technology and scientific research cannot be obtained from nuclear reactors since they are produced only as a result of nuclear reactions involving charged particles or high-energy neutrons. Cyclotrons are used to produce isotopes of this kind as well as certain high-activity isotopes without the use of carriers.

The present work reports on the utilization of a 100-cm cyclotron with a deuteron energy of approximately 10-Mev for the production of widely used isotopes such as Na^{22}, Mn^{52}, Mn^{54}, as well as Zn^{65}.

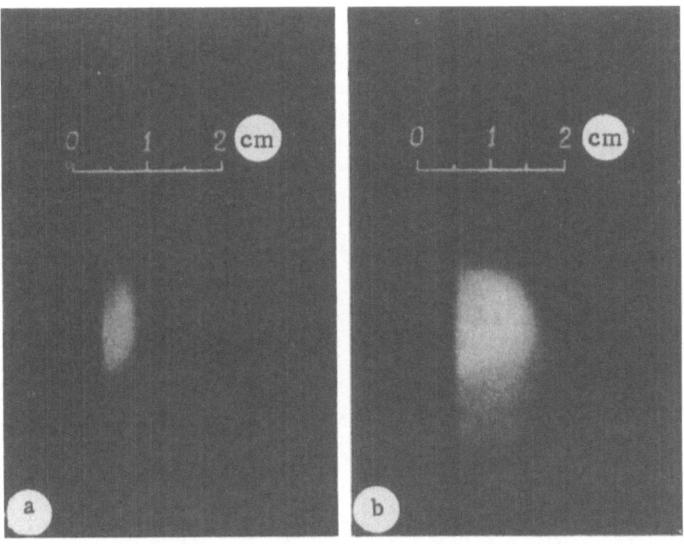

Fig. 1. Radio-autographs of the internal beam at R = 43 cm:
a) minimum ion spread; b) artificially increased ion spread.

In pulsed operation the cyclotron yields a current of 1000 μamp pulsed at the ejection radius of 44 cm. Up to 50 percent of the internal beam is extracted on an external target [1]. The voltage between the dees is 100-105 kv. The maximum average currents achieved in stable operation of the cyclotron are 600-700 μamp for the internal beam and 200-250 μamp for the external beam.

A characteristic difficulty in the use of high ion currents in a cyclotron is the destruction of the target material because of the high current density, especially for the internal beam. The use of the external beam, in which the current density is much smaller, is less advantageous. Furthermore, the passage of high currents through the deflection system leads to instabilities in the deflection voltage, a situation which usually limits the current in the deflected beam to 200-300 μamp.

Data obtained by us using radio-autographs to examine the cross section of the internal beam indicate that at the terminal radius 90% of the beam is distributed over an area of 0.8 cm^2. A power of approximately 1 kw is generated in this area for each one hundred microamperes of beam current.

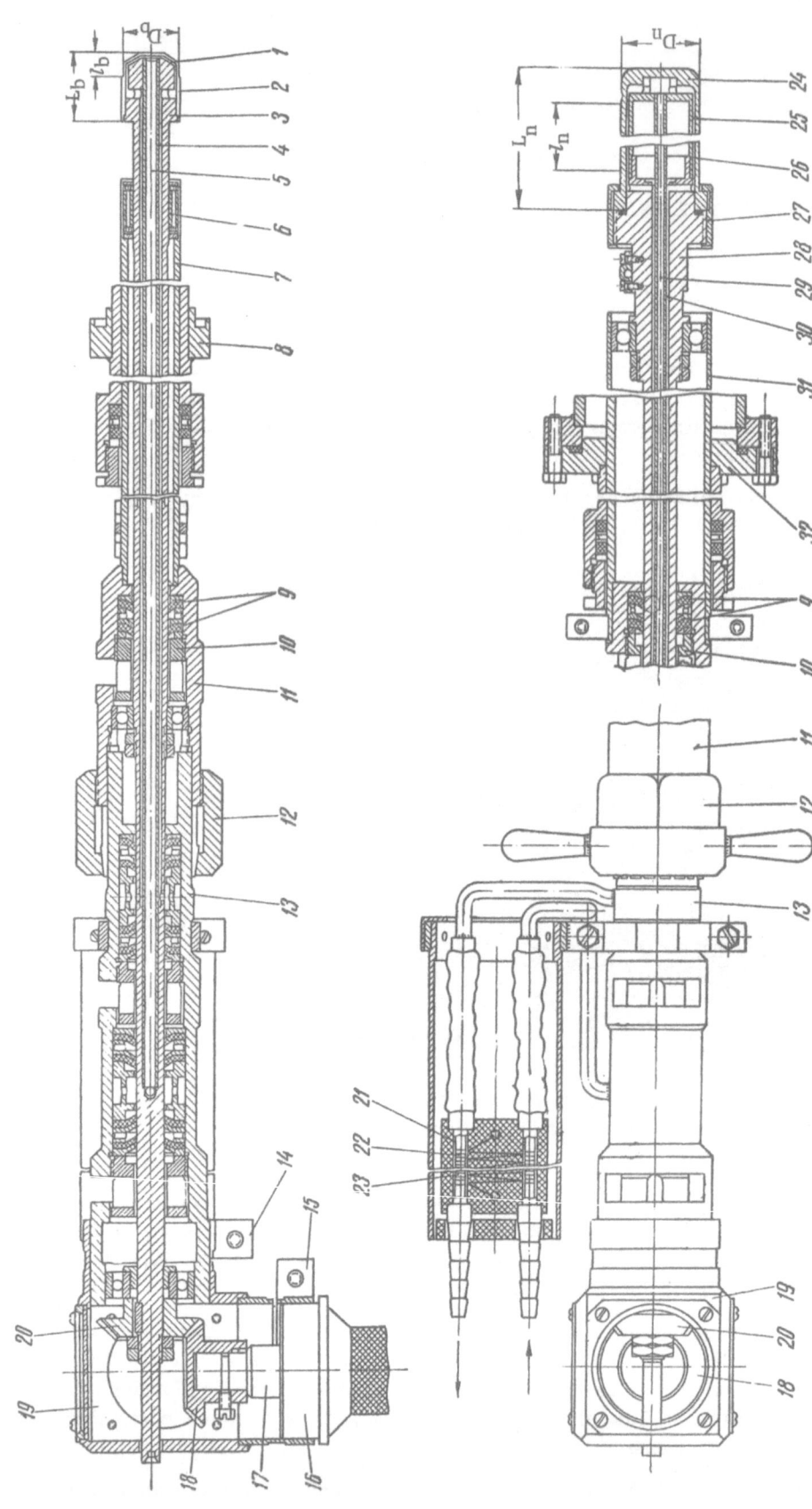

Fig. 2. Diagram of the rotating targets: 1) and 25) gap between the outer surface of the shaft (details 3 and 26) and the inner surface of the cup (details 2 and 24); 2) and 24) bombarded cup; 3) and 28) target shaft; 4) and 30) gap between the outer surface of the copper tube (details 5 and 29) and the inner surface of the target aperture (details 3 and 28); 5) and 29) copper tube; 6) roller bearing; 7) and 31) target rod; 8) and 32) vacuum flange for the target; 9) vacuum gasket; 10), 12) and 27) clamping screws; 11) target frame clamp; 13) target frame; 14) and 15) circular clamps for the angle section; 16) clutch for the flexible shaft; 17) spindle for the flexible shaft; 18) and 20) conical gears; 19) angle section; 21) thermocouple; 22) insulated ring; 23) copper ring supporting thermocouple junction; 26) frame for the shaft head; D_b and D_n — diameter of the bombarded portion of the cup; l_b and l_n — length of the bombarded portion of the cup; L_b and L_n — total length of the cup.

18

To avoid overheating of the bombarded surface it is necessary to improve heat removal and to reduce the current density by increasing the surface being bombarded. This objective is best achieved through the use of a rotating target [2]. In addition, we have explored the possibility of increasing the cross section of the internal beam by exploiting the ion spreading due to the precession of orbit centers. Ion spreading can be increased by displacing the ion source from the position corresponding to minimum ion spread [1], or by using a shim to enhance artificially the first azimuthal harmonic. In Fig. 1 are shown radio-autographs of the beam cross section at a radius of 43 cm for minimum ion spread and artificially increased ion spread. As is apparent from the figure, in the second case the beam cross-sectional area is increased by almost three times. Increasing the ion spread deteriorates the monoenergetic properties of the beam but this is not important when the cyclotron is used to produce radioactive isotopes.

Two rotating targets are used (for the internal and external beams) for producing radioactive isotopes. The construction of the targets is shown in Fig. 2. The target element, which is directly bombarded by the beam, is a rotating metal cup.

A diagram of the target used for irradiation in the internal beam is shown in the upper part of Fig. 2. The metal cup 2 is soft-soldered to the head of the target shaft 3. Inside the shaft there is a coaxial copper tube 5 to provide passage for the cooling water; the exit path for the water is through channel 4, which is formed by the internal surface of the shaft and the outside surface of tube 5. The water which passes through tube 5 and flows out through the circular gap 4 then passes through two thermocouple tubes 21 inside of which are located copper rings 23 which support the thermocouple junction. The thermal electromotive force, which is proportional to the difference in temperatures of the incoming water and outgoing water, is used to determine the beam current which bombards the cup (for a known water flow and deuteron energy). The vacuum seal for the rotating target is made by means of the two rubber rings 9. The water line is sealed by two pairs of rubber gaskets which are located in the target frame 13. The target shaft is held by two ball bearings and one roller bearing, and is rotated by the conical gears 20 and 18, connected by a flexible shaft to the reducer of a 1-kw electric motor. The target is rotated at 112 rpm. The diameter of the bombarded portion of the cup D_b is 28 mm and the wall thickness at this point is 1 mm; the gap 1 through which the cooling water flows is 0.5 mm. The water flow is 5.7 liters per minute at a pressure of 4 atm. This target is used to produce the radioactive isotopes Mn^{54} and Zn^{65}. The cup is made of copper with chrome plate on the external surface in the first case and zinc-free copper (to obtain a high activity) in the second case.

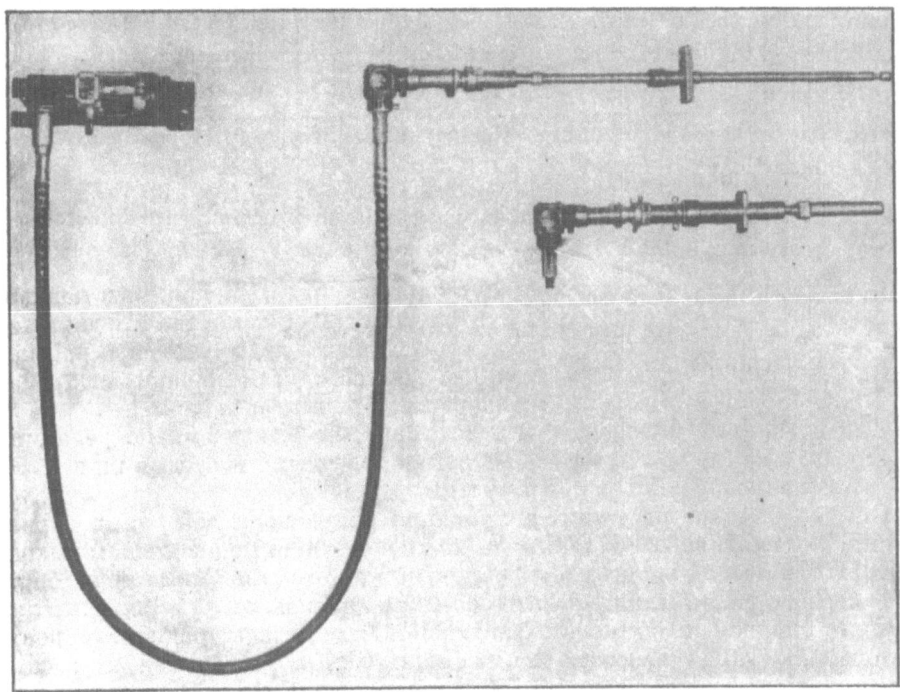

Fig. 3. General view of the rotating targets.

The maximum currents which can be used without destroying the surface of the targets (Fig. 1a) are 250 μ amp for the beam of minimum cross section and 500 μ amp when the beam is used with an artificially increased cross section (Fig. 1b).

In the lower right-hand corner of Fig. 2 is shown a section of the front part of the target used for bombardment by the external beam. The cross section of the external beam at the point at which this target is located is rectangular, with dimensions 180 × 30 cm. The cup 24 is fastened to the head of the target shaft by a screw which squeezes a rubber gasket. The rest of the target is more or less similar to the target used with the internal beam. This target is used to obtain the radioactive isotope Na^{22}. The cup, of diameter D_n = 42 mm, is made from metallic manganese. The wall thickness is 3 mm. The height of the gap 25 for passage of the cooling water is 1 mm. The water flow is 3 liters per minute at a pressure of 4 atm. The target is irradiated by a current of 230 μ amp without any visible deleterious effects to the surface.

In Fig. 3 is shown a general view of both targets; the internal target is shown with the flexible connection to the motor.

The yields of the radioactive isotopes Zn^{65}, Mn^{54} and Na^{22} were determined by comparing the activities of the irradiated targets with the activity of a calibrated Co^{60} sample. The comparison of the activities was carried out with a micro-roentgenmeter of the "Kaktus" type. The values of the ionization constants for the γ-radiation of these isotopes were taken from [6].

The measurements were carried out after the following time periods: Zn^{65} — after 100 days; Mn^{54} — after 85 days, Na^{22} — after 15 days. The yields are shown below.

Isotope	Reaction	Deuteron energy, Mev	Yield, μ Curies /μ amp · hr
Zn^{65} ...	$Cu^{65}(d, 2n)$	10	1,5 $\pm$0,3
Mn^{54}...	$Cr^{53}(d, n)$; $Cr^{54}(d, 2n)$	10	0,65$\pm$0,13
Na^{22}...	$Mg^{24}(d, \alpha)$	10,6	0,87$\pm$0,17

A comparison of these yields with the published data indicates that the Mn^{54} yield is in agreement with the data in [3] within the experimental errors. The Na^{22} yield has been most carefully studied in [4] and [5]. However, the values obtained in these papers differ by a factor of 1.7. The present results are in good agreement with the data reported in [4].

An investigation was made of the use of high-energy neutrons to obtain isotopes for isotopes produced in (n, 2n), (n, α) and (n, p) reactions.

In most cases fast neutrons are obtained by irradiating beryllium or lithium with deuterons. Data on the high yield of high-energy neutrons in the bombardment of a boron target by deuterons have been reported in [7].

In this connection experiments were carried out to study the activation of carbon by neutrons from boron and lithium targets. Measurements were made of the radioactivity of the isotope C^{11}, which is formed in the C^{12} (n, 2n) reaction, and which has a very high energy threshold. However, the C^{11} yield was three or four times smaller when a boron target was used as compared with the lithium target; for this reason a metallic lithium target was used to obtain radioactive isotopes.

LITERATURE CITED

[1] P. P. Dmitriev, N. N. Krasnov and E. N. Khaprov, "The Problem of Beam Deflection in a Cyclotron," J. Atomic Energy (USSR) 7 (1957).

[2] M. S. Livingston, "The Cyclotron," J. Appl. Phys. (1944).

[3] P. Kafalas and J. W. Irvine, Phys. Rev. 104, 703 (1956).

[4] J. W. Irvine and E. T. Clarke, J. Chem. Phys. 16, 686 (1948).

[5] N. A. Vlasov et al., J. Atomic Energy (USSR) 2, 169 (1957).

[6] Bochkarev, Keirim-Markus, L'vova and Pruslin, Measurement of Activity of Sources of Beta- and Gamma-Radiation (AN SSSR Press, 1953).

[7] Bogdanov et al., J. Exptl.-Theoret. Phys. (USSR) 30, 981 (1956).

DETERMINATION OF PRODUCT YIELDS IN NUCLEAR REACTIONS

M. Z. Maksimov

When nuclear reactions are used to obtain radioactive isotopes, it is necessary to have at least a rough idea of the dependence of yield on energy of the incident particles ($d, p, \alpha \ldots$). A knowledge of this dependence then makes it possible to choose a suitable target and to determine which isotopes must be enriched in order to enhance the yield at a given incident-particle energy.

Experimental data on the dependence of yield on energy are scanty. Most data refer to 14-Mev deuterons [1,2]. However, to obtain isotopes one often uses deuterons of other energies and even other particles. Hence it is of interest to ascertain the dependence of yield on energy of the incident particles.

In the present report, we make a rough estimate of the yields of certain isotopes as a function of deuteron energy (up to 20-Mev).

By yield B we are to understand the number of nuclei of a given nuclide produced in a nuclear reaction characterized by a cross section $\sigma(E_a)$ in a thickness $\underline{x}$, when a target is irradiated by a flux of $j_a/(Z_a \cdot e)$ part./$cm^2 \cdot$ sec:

$$B = \frac{j_a}{Z_a \cdot e} \cdot N \int_0^x \sigma(E_a) \cdot dx. \tag{1}$$

Using the relationship for the energy lost by charged particles [3], and expressing j_a in μ amp, $\sigma(E_a)$ in barns, E_a in Mev, and B in units of $3.7 \cdot 10^{10}$, we obtain the following expression for yields from a thick target

$$B \approx \frac{0.7 \cdot j_a}{a \cdot Z_a^3} \cdot N \int_0^{E_a} \frac{\sigma(E_a) E_a \cdot dE_a}{\sum Z_i N_i \ln \frac{200 E_a}{a \cdot Z_i}}, \text{ Curies} \tag{2}$$

where the summation is carried out over all atoms Z_i in the target; Z_a and $\underline{a}$ are the charge and mass number of the incident particle; N is the number of target nuclei in 1 cm^3 (in the case of a mixed target — the number of nuclei from which the given nuclides are produced). The activity expressed in curies per microampere-hour is

$$Q = 0.69 \text{ B}/\text{T} \cdot j_a, \tag{3}$$

where T is the half-life in hours.

A basic difficulty in computing the yield B is a determination of the cross section $\sigma(E_a)$; this cross section can be computed using the statistical model of the nucleus, taking account of the Bohr analysis of nuclear reactions

$$\sigma(E_a) = \sigma_c(E_a) \cdot \eta_B(E_a).$$

Here $\sigma_c(E_a)$ is the cross section for the formation of the compound nucleus; $\eta_B(E_a)$ is the decay probability for the compound nucleus with subsequent formation of a given nuclide. It is assumed that the energy distribution of the emitted particles $\underline{b}$ is of the form [4]

$$I(E_b) dE_b \sim b \cdot g_b \cdot (2I_B + 1) \cdot \sigma_c(E_b) \cdot e^{-\frac{E_b - \varepsilon_b}{\tau}} dE_b, \tag{4}$$

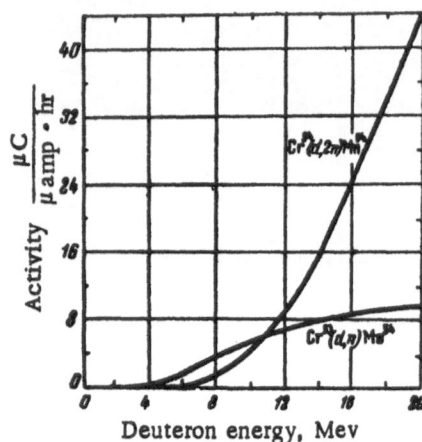

Fig. 1. Activity of Mn^{54} as a function of deuteron energy, computed for 100% content of Cr^{53} and Cr^{54}, respectively.

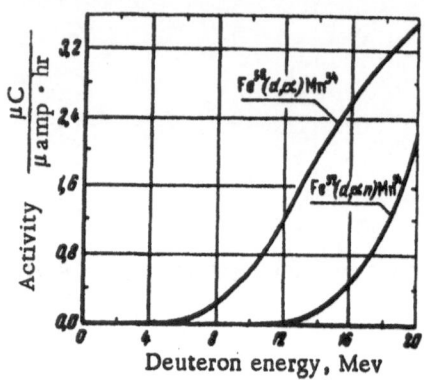

Fig. 2. Activity of Mn^{54} as a function of deuteron energy, computed for 100% content of Fe^{56} and Fe^{57}, respectively.

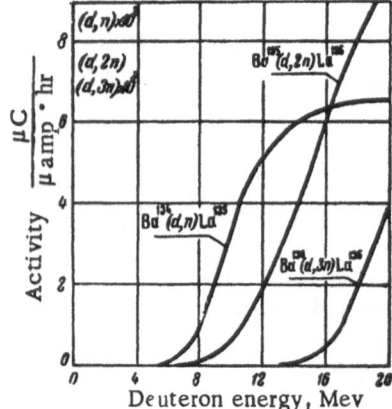

Fig. 3. Activity of La^{135} as a function of deuteron energy, computed for 100% content of Ba^{134}, Ba^{135}, and Ba^{136}, respectively.

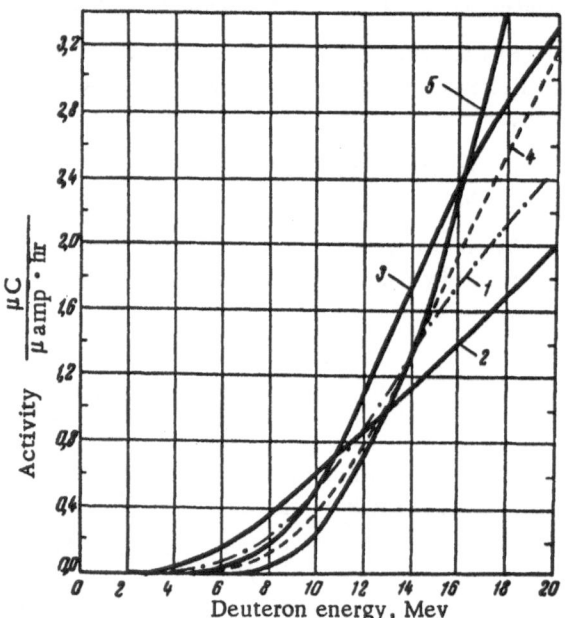

Fig. 4. Activity as a function of deuteron energy: 1) Na^{22} from Mg; 2) Mn^{54} from Cr; 3) Mn^{54} from Fe; 4) Zn^{65} from Cu, 1:2; 5) La^{135} from Ba, 1:200.

where $\underline{b}$ and g_b are the mass number and statistical weight of the emitted particle; I_B is the spin of the residual nucleus; $\sigma_c(E_b)$ is the cross section for the formation of the compound nucleus in the inverse process; ϵ_b is the binding energy of the particle $\underline{b}$ in this nucleus and τ is the mean "temperature" of the compound nucleus, the dependence of which on the energy of the incident particle is of the form [5]

$$\frac{1}{\tau} \approx \frac{\pi}{2\sqrt{33}} \left\{ \frac{A+a}{E_a - \epsilon_a} \right\}^{1/3}, \quad \text{Mev}^{-1}. \tag{5}$$

Here A + a is the mass number of the compound nucleus and ϵ_a is the binding energy of the incident particle in this nucleus.

The interaction radii of particles with nuclides which appear in the cross sections $\sigma_c(E_a)$ and $\sigma_c(E_b)$ are taken in the following form

$$R_p = 1.5 \cdot A^{1/3} \cdot 10^{-13}, \quad \text{cm} \qquad R_d = R_\alpha = (1.5 \cdot A^{1/3} + 1.2) \cdot 10^{-13}, \quad \text{cm}$$

Details of the calculation of $\sigma_c(E_a)$, $\sigma_c(E_b)$ and $\eta_B(E_a)$ are given in [4-6] and [10].

In this way we have computed the yields for Na[22] (2.6-year), Mn[54] (300-day), Zn[65] (250-day), La[135] (19.5-hour) as a function of deuteron energy in the following reactions:

$$Mg^{24;\,25}\,(d,\,\alpha;\,\alpha n)\,Na^{22}; \quad Cr^{53;\,54}\,(d,\,n;\,2n)\,Mn^{54}; \quad Fe^{56;\,57}\,(d,\,\alpha;\,\alpha n)\,Mn^{54};$$
$$Cu^{65}\,(d,\,2n)\,Zn^{65}; \quad Ba^{134;\,135;\,136}\,(d,\,n;\,2n;\,3n)\,La^{135}.$$

In these calculations no account has been taken of reactions which occur without the formation of a compound nucleus (stripping and separation of the deuteron in the Coulomb field of the nucleus, etc.); first, because of the fact that these processes take place for the (d, n) and (d, p) reactions; second, because the effect on the yield of various nuclides is less important than on the angular distribution of neutrons or protons; and third, because a calculation to take these processes into account is difficult because of the absence of quantitative data.

The results of the calculations are shown in Figs. 1-4; in Figs. 1-3 is shown the dependence of yield (activity) for one nuclide, while in Fig. 4 account is taken of the percentage content of the corresponding nuclides in the target. In Table 1 a comparison is made between the experimental and theoretical data.

TABLE 1

E_d, Mev	Na[22] from Mg		Mn[54] from Cr		Mn[54] from Fe		Zn[65] from Cu	
	Experimental data	Theoretical data	Experimental data	Theoretical data	Experimental data	Theoretical data	Experimental data	Theoretical data
8	0.43*	0.23	0.45*	0.36	0.10**	0.19	—	0.23
10.5	0.97*	0.60	0.75*	0.63	—	0.65	—	0.90
14	1.77*	1.35	1.15*	1.13	1.00**	1.73	3.4*	2.63
14	1.00**	—	—	—	—	—	0.5**	—

*Data of [1, 9].
**Data of [2, 8].

It is apparent from Table 1 that the discrepancies between the experimental and theoretical data are much greater than the accuracy of the measurements. The theoretical data may differ from the experimental data for two reasons: first, because of the fact that an approximation has been used in the calculations [Eqs. (2), (4) and (5)], and second, because of the fact that the individual properties of the nuclides have been taken into account only by the factor $(2I_B + 1)$ and the binding energy ϵ_b. Apparently this procedure is not sufficient for describing the individual features of even-even, even-odd, and other nuclides. However, it is apparent from Table 1 and from the calculations that these factors do not play a very important role in the energy dependence of the yields. Hence, the curves shown in Figs. 1-4 are useful in the sense that they can be utilized to choose a suitable target and to determine which isotopes of the target should be enriched in order to enhance the yield at a given incident-particle energy.

For example, a comparison of curves 2 and 3 in Fig. 4 shows that when $E_d > 10$ Mev it is desirable to use an Fe target to obtain the highest Mn[54] yield; similarly, a Cr target is indicated when $E_d < 10$ Mev. On the other hand, an examination of the curves in Fig. 1 shows that when $E_d > 10$ Mev, in order to increase the Mn[54] yield it is necessary to enrich the chromium target in the Cr[54] isotope; similarly, one should enrich the Cr[53] isotope when $E_d < 10$ Mev. A more detailed analysis of the relations by which these curves were obtained indicates that the greatest Mn[54] yield is obtained from a thick target, in which the energy is degraded from 8 Mev to 4 Mev, by enriching the Cr[53] isotope; in a target where 8 Mev is lost, it is desirable to enrich the Cr[54] isotope. This isotopic distribution yields a stratified enrichment (the appropriate thicknesses can be computed from the range-energy curve [7]).

In a similar way, one can estimate the product yields in other reactions involving bombardment by protons and α-particles and carry out a similar analysis to obtain the highest yield for a given isotope. The calculations indicate that [cf. Eq. (2)] to obtain the maximum yield from thick targets it is desirable to use protons rather than deuterons and, of course, α-particles. This situation arises from the fact that the interaction cross sections

$\sigma_c(E_a)$ for protons and deuterons of the same energy are approximately the same, whereas the energy loss is smaller for protons than for deuterons. Consequently, the "effective" range (activation thickness) is greater for protons. In this case, there are competing reactions which lead to the same final product: for example, $(p, \gamma) - (d, n)$; $(p, n) - (d, 2n)$; $(p, 2n) - (d, 3n)$, etc. Although the (p, γ) reaction is much weaker than the (d, n) reaction in the case of monoisotopic targets, the (p, n), the $(p, 2n)$, etc., reactions are stronger than the $(d, 2n)$, $(d, 3n)$, etc., since the thresholds for the latter are higher.

On the other hand, if the same machine (cyclotron) is used for particle acceleration, for given machine parameters, R and H, we have:

$$E_p = E_\alpha = 2E_d = 4.8 \cdot 10^{-5} (H \cdot R)^2 \text{ Mev}, \qquad (6)$$

where H is the magnetic field in kilogauss and R is the radius in cm.

Consequently, in this case the "effective" proton range in the material will be still greater ($E_p = 2E_d$!). Furthermore, the excitation energy of the nuclei will be higher for protons than for deuterons if $E_d > |\epsilon_d| - |\epsilon_p| \approx 5$ Mev ($\epsilon_d \approx -13$ Mev; $\epsilon_p \approx -8$ Mev). This condition can be expressed in terms of the machine parameters;

$$H \cdot R > 400\text{--}450 \text{ kilogauss} \cdot \text{cm}. \qquad (7)$$

For comparison purposes, in Table 2 are shown the calculated yields of Zn^{65} in $\dfrac{\mu C}{\mu \, \text{amp} \cdot \text{hour}}$ for thick targets using the (p, n) and $(d, 2n)$ reaction in Cu and the excitation energy U of the corresponding residual nuclei Zn^{66} and Zn^{67} as functions of $H \cdot R$.

TABLE 2

	$H \cdot R$ kilogauss cm			
	400	600	800	900
$Cu^{65}(p, n) Zn^{65}$ . . .	1,4	8,8	10,2	10,4
$Cu^{65}(d, 2n) Zn^{65}$ . . .	0,0	0,4	2,9	5,9
$U_p Zn^{66}$, Mev	16,6	26,1	39,1	47,4
$U_d Zn^{67}$, Mev	17,4	22,0	28,4	32,5

Thus, it is apparent that the use of accelerated protons in a cyclotron may be more desirable from the point of view of increasing yield, as well as increasing the excitation energy of the nuclei ($H \cdot R > 400$).

LITERATURE CITED

[1] E. Clark and J. Irvine, Phys. Rev. 70, 893 (1946); J. Chem. Phys. 16, 686 (1948).

[2] A. N. Nesmeianov, A. V. Lapitskii and N. P. Rudenko, Production of Radioactive Isotopes, Goskhimizdat, 1954.

[3] E. Segre, Experimental Nuclear Physics, Vol. I, Part 2 [Russian translation] (IL, 1956).

[4] J. Blatt and V. Weisskopf, Theoretical Nuclear Physics [Russian translation] (IL, 1954).

[5] K. J. LeCouteur, Proc. Phys. Soc. 63A, 259 (1950).

[6] S. N. Choshal and T. N. Dave, Indian J. Phys. 24, N. 4, 213 (1953).

[7] B. V. Rubikov, J. Exptl.-Theoret. Phys. (USSR) 28, 651 (1955).

[8] G. Hevesey, Radioactive Indicators [Russian translation] (IL, 1950).

[9] P. Kafalas and J. Irvine, Phys. Rev. 104, 703 (1956).

[10] M. Z. Maksimov, J. Exptl.-Theoret. Phys. (USSR) 33, No. 12 (1957).

SPECTROCHEMICAL METHODS OF ANALYZING HIGH-PURITY MATERIALS USED IN REACTOR CONSTRUCTION AND FOR THE PRODUCTION OF RADIOISOTOPES*

A. G. Karabash and Sh. I. Peizulaev

A large assortment of pure metals and their compounds are used in modern technology, and the purity requirements for them continually increase.

The usefulness of artificial radioactive isotopes for scientific and industrial investigations is determined to a considerable degree by the high purity of the starting materials from which they are prepared. The presence of traces of impurities in the materials, in particular elements with high activation cross sections, which form long-lived radioisotopes, can lead to a series of difficulties or errors in subsequent work.

For control of the quality of raw materials and intermediate products in the production of radioactive isotopes, the most important and direct method is radioactivation analysis [1-5]. It has the maximum sensitivity and therefore offers exceptional possibilities of determining micro amounts of impurities. However, the application of this method to complex combinations of elements has not been sufficiently developed as yet. The bulk of the data available in the literature characterizes only the absolute sensitivity of the radioactivation determination of separate elements, without considering the effect of interfering components and the necessity of a preliminary chemical separation [2, 3, 5 and 6].

Together with radiochemical methods, in the control of radioisotope production it is also expedient to use spectral, chemical, and physicochemical methods of determining elements present as traces. They are well developed and play an important role in the control of high-purity products [7-10]. In the modern analysis of traces of metals, an exceptional position is occupied by emission spectral analysis due to the high absolute sensitivity and the possibility of simultaneously determining elements in the presence of each other. Spectral methods are used for determining a large number of elements (about 70) in the most varied products [11-16]. However, problems are often encountered where both chemical and direct spectral methods are insufficiently sensitive. In such cases a preliminary enrichment of the impurities is applied in spectral analysis.

The known method of evaporating impurities, developed by a group of Soviet authors [17, 18], is based on the principle of physical enrichment.

A combination of chemical enrichment of the impurities with subsequent spectral determination has already been used for a long time in the analysis of traces in the form of so-called combination methods both by Soviet and foreign investigators [19-35].

We should note that most of the problems solved by combination methods are concerned with the field of analyzing biological materials, rock, and other products with a complex composition. Here the trace impurities are usually separated from the bulk with selective or group reagents.

Another type of problem is presented by the analytical control of pure metals and their compounds in which the macrocomponent is one element — the base — and the microcomponents — impurities — are present in large numbers and are to be determined.

---

* The methods were developed under the direction of A. G. Karabash and Sh. I. Peizulaev, with the participation of R. L. Sliusareva, Z. N. Samsonova, L. S. Krauz, N. P. Sotnikova, N. I. Smirnova-Averina, G. G. Morozova, L. S. Romanovich, I. I. Smirenkina, V. M. Lipatova, S. K. Sazanova, V. P. Usacheva, L. I. Pugacheva, F. A. Kostareva, P.D. Gorbachev, E.F. Voronova, F.L. Babina, V.S. Moleva, N.T. Kostereva,N.N. Kuznetsova and A. I. Elovatskaia.

TABLE 1

Analysis of Metals with Simple Spectra

Experiment number	Metal analyzed, or its compound	Chemical preparation		Base of concentrate of impurities and unenriched sample
		Method of separating base in concentration of impurities		
1	Sodium	Precipitation of chloride from hydrochloric acid solution with gaseous HCl		NaCl
2	Potassium	The same		KCl
3	Aluminum	" "		Al_2O_3
4	Calcium	Precipitation of sulfate with sulfuric acid		$CaSO_4$
5	Strontium	The same		$SrSO_4$
6	Barium	" "		$BaSO_4$
7	Lead	" "		$PbSO_4$
8	Bismuth	Precipitation of the iodide BiI_3 with hydriodic acid		Bi_2O_3
9	Gallium	Ether extraction from a solution of the chlorides in 6N hydrochloric acid		BeO, Ga_2O_3
10	Germanium	Distillation of the chloride from hydrochloric acid solution		GeO_3
11a	Tin	The same (in the presence of H_2O_2)		SnO_2
11b		Chlorination of the metal with gaseous chlorine		SnO_2
12	Silicon	Volatilization of the fluoride (by treatment with a mixture of hydrofluoric, nitric and sulfuric acids)		$SrSO_4$
13	Gold	Reduction to the metal with hydrogen sulfide in hydrochloric acid solution		BeO
14	Magnesium	Extraction of the impurities with organic solvents with dithizone and hydroxyquinoline		$MgSO_4$
15a	Alkali metals (sodium, etc.)	The same		BeO, etc.
15b		Coprecipitation of the impurities with hydroxyquinoline		

Enrichment factor of impurities in concentrate	Spectral analysis			
	Impurity elements			
	determined simultaneously		determined separately	total number
	in concentrate	only in sample without enrichment		
20-30	Mg, Ba, Al, Ti, Cr, Mo, Fe, Co, Ni, Cu, Cd, Sn, Sb	Ca, Mn, Ag, Pb, Bi	B, Li, K	21
20-30	Mg, Ba, Al, Ti, Cr, Mo, Mn, Fe, Co, Ni, Cu, Ag, Cd, Sn, Sb	Ca, Pb, Bi	B, Li, Na	21
40-50	Mg, Ca, Ba, Ti, V, Cr, Bi, Mo, Mn, Fe, Co, Ni, Cu, Ag, Zn, Cd, Sn, Pb, Sb	Be, Au, In, Te		23
25-60	Mg, Al, Ti, Cr, Mn, Fe, Co, Ni, Cu, Ag, Zn, Cd, Sn	Ba, Pb		15
20-30	Mg, Ca, Al, Ti, Cr, Mn, Fe, Co, Ni, Cu, Ag, Zn, Cd, In, Sn, Sb, Bi	Ba, Pb	Na	20
~30	Be, Mg, Ca, Al, Ti, Cr, Mn, Fe, Co, Ni, Cu, Zn, Cd, Sn, Sb	Ag, Pb, Bi		18
30-60	Mg, Ca, Al, Ti, V, Cr, Mo, Mn, Fe, Co, Ni, Cu, Ag, Zn, Cd, In, Tl, Sn, Sb, Bi, As	Ba, Pt, Au, Te	B, Li, Na	28
25-30	Mg, Ca, Ba, Al, Ti, V, Cr, Mo, Mn, Fe, Co, Ni, Zn, Cd, In, Sb, Te, As	Pt, Cu, Ag, Au, Tl, Sn, Pb	B, Li, Na	28
50	Mg, Ca, Ba, Al, Ti, V, Cr, Mn, Co, Ni, Cu, Ag, Zn, Cd, Pb, Bi			16
50-100	Mg, Ca, Ba, Al, Ti, V, Cr, Mo, Mn, Fe, Co, Ni, Cu, Ag, Au, Zn, Cd, In, Tl, Sn, Pb, Sb, Bi			23
15-20	Mg, Ca, Ba, Al, Ti, V, Cr, Mo, Mn, Fe, Co, Ni, Cu, Ag, Zn, Cd, In, Pb, Bi	Sb		20
15-30	Mg, Ca, Ba, Al, Cr, Mn, Fe, Co, Ni, Cu, Ag, Zn, Cd, In, Pb, Bi	Ti, V, Mo, Sb		20
20-30	Mg, Ca, Ba, Al, V, Ti, Cr, Mn, Fe, Co, Ni, Cu, Ag, Zn, Cd, Sn, Pb, Sb, Bi		Na	20
10	Mg, Ca, Ba, Al, Ti, V, Cr, Mo, Mn, Fe, Co, Ni, Cu, Ag, Zn, Cd, Sn, Pb, Bi, Sb			20
~50	Al, Ti, V, Mn, Fe, Co, Ni, Cu, Ag, Cd, Sn, Pb, Bi	Ca, Ba, Cr, Sb		17
50-100	Al, Ti, V, Mo, Mn, Fe, Co, Ni, Cu, Ag, Zn, Cd, Sn, Pb, Bi	Mg, Ca, Ba, Cr		19
50-100	Mg, Al, Ti, V, Mn, Fe, Co, Ni, Cu, Ag, Cd, In, Pb, Bi			14

TABLE 2

Analysis of Metals with Complex Spectra

Expt. No.	Metal analyzed or its compound	Chemical Preparation — Method of separating base in concentration of impurities	Base of concentrate of impurities and unenriched sample	Enrichment factor of impurities in concentrate	Spectral analysis — Impurity elements determined simultaneously		determined separately	total number
					in concentrate	only in sample without enrichment		
16a	Molybdenum	Ether extraction from solution of chlorides in 6N hydrochloric acid	MoO_3+C+Ga_2O_3	25–50	Mg, Ca, Ba, Al, Cr, Mn, Co, Ni, Cu, Ag, Cd, Pb, Bi	Fe, Sb	Si	16
16b		Flotation in chloroform of precipitate formed with α-benzoinoxime	BeO	5–10	Mg, Ca, Ba, Al, Ti, V, Cr, Mn, Fe, Co, Ni, Cu, Ag, Zn, Cd, In, Sn, Pb, Sb, Bi		Na	21
17	Iron	Ether extraction from solution of chlorides in 6N hydrochloric acid	BeO	50–100	Mg, Ca, Ba, Al, Ti, V, Cr, Mn, Co, Ni, Cu, Ag, Zn, Cd, Pb, Bi			16
18	Nickel	Chloroform flotation of precipitate with dimethylglyoxime	BeO	3	Mg, Ca, Ba, Al, Ti, V, Cr, Mo, Mn, Fe, Co, Cu, Ag, Zn, Sn, Pb			16
19	Chromium	Volatilization of CrO_2Cl_2 from solution in hydrochloric and perchloric acids	$SrSO_4$	10–20	Be, Mg, Ca, Al, Ti, Mo, Mn, Fe, Ni, Cu, Co, Ag, Zn, Cd, In, Pb, Bi, Tl, Ga		Na	20

For application to this problem we developed new "spectrochemical" methods for analyzing pure metals and their compounds for impurity content. This name is convenient for such a combination of two methods.

The methods developed are based on a general principle in which a large number of impurities are concentrated by selective separation of the base element and all the impurities in the concentrate are subsequently determined spectrographically. In the analysis methods developed for various metals, listed in Tables 1 and 2, there is a series of general procedures both in the chemical preparation and in the spectral analysis.

The impurities are concentrated by separating the base element by conversion into another phase, solid, liquid, or gas.

a) Separation in the solid phase. The base element is converted into an insoluble compound and separated as a precipitate from the mother solution by decantation, and in some cases by centrifuging (the analysis of sodium, potassium, aluminum, calcium, strontium, barium, lead, bismuth and gold).

b) Separation in the liquid phase. The base element is extracted as a low-polarity compound from an aqueous solution with an organic solvent that is immiscible with it (the analysis of gallium, iron and molybdenum). Use was also made of the quantitative flotation of precipitates of metal complexes with the help of organic liquids [36] (the analysis of molybdenum and nickel).

c) Separation in the gas phase. The base element is removed in the form of a gaseous compound or distilled off from an aqueous solution (the analysis of silicon, germanium, tin and chromium) or as a result of volatilization by the direct action of a gaseous reagent on the sample in a solid form (chlorination in the analysis of tin). In most cases, the separation of only the bulk of the base element was required in the chemical concentration and not the absolutely complete removal of it from the concentrate of impurities. Therefore, it was possible to use many selective reagents, which are usually not applicable in quantitative analysis.

The highly selective separation of the base element may be illustrated by the following example. In the periodic table, a small circle marks those elements, traces of which can be 90-100% concentrated in an acid solution in the separation of molybdenum as the complex with α-benzoinoxime by quantitative flotation of the precipitate with chloroform.

The selective reactions and conditions were chosen so that the impurities were concentrated quantitatively, reliably, simply and, if possible, in one operation. The reagents used were such that the excess could be removed from the concentrate readily by volatilization, oxidation, or thermal decomposition. The preferred reagents were the simplest to prepare in a spectrally pure form by distillation, sublimation, etc.

The chemical operations were performed predominantly in quartz vessels.

A blank experiment was performed to give the quantitative correction for the impurities contained in the reagents and the vessel.

The bulk of the selective reagents used, shown in Tables 1 and 2, have been used for a long time in classical chemical analysis for the separation of elements, and studied thoroughly. However, there were frequently insufficiently full literature data on the separation of microconcentrations of the elements. In addition, in most of the methods the experimental conditions of the classical reactions were substantially changed. Thus, the separation of crystalline precipitates of the base element, for example barium, strontium, calcium, and lead sulfates, aluminum chloride, etc., was carried out as a rule under conditions corresponding to high solubility of the precipitate, which favored recrystallization and self-purification of the precipitates. Therefore, it was necessary to test the applicability of the reactions in detail for the quantitative separation of traces of elements on artificial mixtures of salts, in which the concentrations of the added elements were varied from comparatively large ($\sim 10^{-1}\%$) to minimal (10^{-4}-$10^{-6}\%$). In each case an investigation was made of the distribution of traces of a large number of elements between phases under the analysis conditions to verify that the impurity elements were completely extracted into the concentrate. For this purpose we also used radioactive tracers and activation analysis. As a result we determined the range of elements, traces of which were concentrated practically quantitatively, i.e., $\sim 90\%$ and more (see Tables 1, 2 and 3). In most cases of base element separation the behavior of microconcentrations of impurities corresponded to the well-known behavior of macroconcentrations of these elements, i.e., the distribution of the impurities between phases remained practically constant over a wide range of concentrations. The exceptions observed were very few.

TABLE 3

Sensitivity of Spectrochemical Methods of Determining Impurity Elements (in %)

Metal analyzed (or its compound)	Ref. to characteristics of method in Tables 1, and 2**	Li	Na*	Be	Mg*	Ca*	Ba	Al*	Ti
1	2	3	4	5	6	7	8	9	10
Sodium	1 15a	$3 \cdot 10^{-5}$			$5 \cdot 10^{-5}$	$5 \cdot 10^{-8}$	$3 \cdot 10^{-4}$	$5 \cdot 10^{-5}$ $4 \cdot 10^{-5}$	$3 \cdot 10^{-4}$ $4 \cdot 10^{-5}$
Potassium	2	$3 \cdot 10^{-5}$	$3 \cdot 10^{-4}$		$5 \cdot 10^{-5}$	$5 \cdot 10^{-8}$	$3 \cdot 10^{-4}$	$5 \cdot 10^{-5}$	$3 \cdot 10^{-4}$
Aluminum	3			$2 \cdot 10^{-3}$	$4 \cdot 10^{-5}$	$2 \cdot 10^{-4}$	$4 \cdot 10^{-4}$		$2 \cdot 10^{-4}$
Calcium	4		$2 \cdot 10^{-4}$		$1 \cdot 10^{-4}$		$1 \cdot 10^{-3}$	$1 \cdot 10^{-4}$	$6 \cdot 10^{-5}$
Strontium	5		$2 \cdot 10^{-4}$		$3 \cdot 10^{-5}$	$2 \cdot 10^{-4}$	$5 \cdot 10^{-3}$*	$2 \cdot 10^{-4}$	$3 \cdot 10^{-5}$
Barium	6			$1 \cdot 10^{-6}$	$3 \cdot 10^{-5}$	$3 \cdot 10^{-5}$		$3 \cdot 10^{-5}$	$3 \cdot 10^{-5}$
Lead	7	$3 \cdot 10^{-5}$	$1 \cdot 10^{-4}$		$3 \cdot 10^{-5}$	$3 \cdot 10^{-5}$	$1 \cdot 10^{-3}$	$3 \cdot 10^{-5}$	$3 \cdot 10^{-5}$
Bismuth	8	$3 \cdot 10^{-5}$	$3 \cdot 10^{-5}$		$3 \cdot 10^{-5}$	$5 \cdot 10^{-5}$	$5 \cdot 10^{-6}$	$2 \cdot 10^{-4}$	$5 \cdot 10^{-5}$
Gallium	9				$5 \cdot 10^{-5}$	$1 \cdot 10^{-4}$	$6 \cdot 10^{-5}$	$4 \cdot 10^{-5}$	$4 \cdot 10^{-5}$
Germanium	10				$5 \cdot 10^{-5}$	$1 \cdot 10^{-4}$	$1 \cdot 10^{-4}$	$2 \cdot 10^{-5}$	$1 \cdot 10^{-4}$
Tin	11a, 6				$3 \cdot 10^{-5}$	$3 \cdot 10^{-5}$	$1 \cdot 10^{-3}$	$2 \cdot 10^{-5}$	$5 \cdot 10^{-5}$
Silicon	12		$1 \cdot 10^{-4}$		$1 \cdot 10^{-4}$	$5 \cdot 10^{-4}$		$3 \cdot 10^{-4}$	$5 \cdot 10^{-5}$
Gold	13				$3 \cdot 10^{-4}$	$2 \cdot 10^{-3}$	$3 \cdot 10^{-3}$	$2 \cdot 10^{-3}$	$2 \cdot 10^{-3}$
Magnesium	14					$1 \cdot 10^{-3}$	$2 \cdot 10^{-2}$	$3 \cdot 10^{-4}$	$2 \cdot 10^{-5}$
Molybdenum	166 16a		$4 \cdot 10^{-4}$		$6 \cdot 10^{-5}$ $5 \cdot 10^{-5}$	$4 \cdot 10^{-4}$ $6 \cdot 10^{-5}$	$6 \cdot 10^{-4}$ $6 \cdot 10^{-5}$	$4 \cdot 10^{-4}$ $5 \cdot 10^{-5}$	$6 \cdot 10^{-4}$
Iron	17				$3 \cdot 10^{-5}$	$4 \cdot 10^{-5}$	$6 \cdot 10^{-5}$	$4 \cdot 10^{-5}$	$4 \cdot 10^{-5}$
Nickel	18				$1 \cdot 10^{-4}$	$1 \cdot 10^{-3}$	$1 \cdot 10^{-3}$	$1 \cdot 10^{-3}$	$1 \cdot 10^{-3}$
Chromium	19			$1 \cdot 10^{-5}$	$1 \cdot 10^{-4}$	$5 \cdot 10^{-4}$		$3 \cdot 10^{-4}$	$5 \cdot 10^{-5}$

* The sensitivity shown takes into account the blank experiment.

** This column gives the index number of the given method in Tables 1 and 2.

V	Cr.	Mo	Mn	Fe*	Co	Ni	Cu	Ag	Au
11	12	13	14	15	16	17	18	19	20
	$5\cdot10^{-5}$	$3\cdot10^{-4}$	$3\cdot10^{-4}$	$5\cdot10^{-5}$	$2\cdot10^{-4}$	$2\cdot10^{-4}$	$2\cdot10^{-5}$	$3\cdot10^{-4}$	
$2\cdot10^{-5}$	$4\cdot10^{-5}$	$4\cdot10^{-5}$	$4\cdot10^{-6}$	$2\cdot10^{-5}$	$4\cdot10^{-5}$	$4\cdot10^{-5}$	$4\cdot10^{-6}$	$6\cdot10^{-6}$	
	$5\cdot10^{-5}$	$3\cdot10^{-4}$	$2\cdot10^{-5}$	$5\cdot10^{-5}$	$2\cdot10^{-4}$	$2\cdot10^{-4}$	$2\cdot10^{-5}$	$1\cdot10^{-5}$	
$6\cdot10^{-5}$	$4\cdot10^{-5}$	$4\cdot10^{-5}$	$4\cdot10^{-6}$	$4\cdot10^{-5}$	$6\cdot10^{-5}$	$4\cdot10^{-5}$	$4\cdot10^{-5}$	$4\cdot10^{-6}$	$4\cdot10^{-4}$
	$2\cdot10^{-5}$		$1\cdot10^{-5}$	$4\cdot10^{-5}$	$2\cdot10^{-5}$	$6\cdot10^{-5}$	$6\cdot10^{-6}$	$6\cdot10^{-6}$	
	$5\cdot10^{-4}$		$3\cdot10^{-6}$	$2\cdot10^{-4}$	$1\cdot10^{-4}$	$2\cdot10^{-4}$	$2\cdot10^{-5}$	$2\cdot10^{-5}$	
	$3\cdot10^{-5}$		$3\cdot10^{-6}$	$3\cdot10^{-5}$	$1\cdot10^{-4}$	$1\cdot10^{-4}$	$3\cdot10^{-6}$	$3\cdot10^{-4}$	
$6\cdot10^{-6}$	$1\cdot10^{-5}$	$3\cdot10^{-5}$	$1\cdot10^{-6}$	$3\cdot10^{-5}$	$2\cdot10^{-5}$	$3\cdot10^{-5}$	$3\cdot10^{-6}$	$1\cdot10^{-6}$	$1\cdot10^{-3}$
$1\cdot10^{-5}$	$2\cdot10^{-5}$	$5\cdot10^{-5}$	$2\cdot10^{-6}$	$3\cdot10^{-5}$	$2\cdot10^{-5}$	$5\cdot10^{-5}$	$2\cdot10^{-5}$	$2\cdot10^{-5}$	$6\cdot10^{-5}$
$2\cdot10^{-5}$	$2\cdot10^{-5}$		$4\cdot10^{-6}$		$4\cdot10^{-5}$	$4\cdot10^{-5}$	$4\cdot10^{-6}$	$6\cdot10^{-6}$	
$2\cdot10^{-5}$	$2\cdot10^{-5}$	$2\cdot10^{-4}$	$1\cdot10^{-5}$	$2\cdot10^{-4}$	$1\cdot10^{-4}$	$1\cdot10^{-4}$	$2\cdot10^{-5}$	$1\cdot10^{-6}$	$2\cdot10^{-5}$
$5\cdot10^{-5}$	$2\cdot10^{-5}$	$5\cdot10^{-4}$	$2\cdot10^{-5}$	$2\cdot10^{-4}$	$5\cdot10^{-5}$	$5\cdot10^{-5}$	$1\cdot10^{-5}$	$4\cdot10^{-6}$	
$5\cdot10^{-4}$	$3\cdot10^{-4}$	$5\cdot10^{-5}$	$1\cdot10^{-5}$	$5\cdot10^{-4}$	$3\cdot10^{-4}$	$2\cdot10^{-4}$	$5\cdot10^{-5}$	$2\cdot10^{-5}$	
$1\cdot10^{-3}$	$1\cdot10^{-3}$	$2\cdot10^{-3}$	$2\cdot10^{-4}$	$1\cdot10^{-3}$	$2\cdot10^{-3}$	$2\cdot10^{-3}$	$2\cdot10^{-4}$	$3\cdot10^{-4}$	
$4\cdot10^{-5}$	$2\cdot10^{-3}$		$1\cdot10^{-5}$	$3\cdot10^{-4}$	$6\cdot10^{-5}$	$4\cdot10^{-5}$	$6\cdot10^{-6}$	$6\cdot10^{-6}$	
	$2\cdot10^{-4}$		$4\cdot10^{-5}$	$2\cdot10^{-4}$	$4\cdot10^{-4}$	$4\cdot10^{-4}$	$4\cdot10^{-5}$	$6\cdot10^{-5}$	
$2\cdot10^{-4}$	$6\cdot10^{-5}$		$2\cdot10^{-6}$	$1\cdot10^{-3}$	$6\cdot10^{-5}$	$6\cdot10^{-5}$	$2\cdot10^{-6}$	$6\cdot10^{-6}$	
$2\cdot10^{-5}$	$2\cdot10^{-5}$		$4\cdot10^{-6}$		$4\cdot10^{-5}$	$4\cdot10^{-5}$	$4\cdot10^{-6}$	$6\cdot10^{-6}$	
$3\cdot10^{-4}$	$3\cdot10^{-4}$	$1\cdot10^{-3}$	$1\cdot10^{-4}$	$3\cdot10^{-4}$	$1\cdot10^{-3}$		$1\cdot10^{-4}$	$1\cdot10^{-4}$	
$5\cdot10^{-4}$		$5\cdot10^{-5}$	$1\cdot10^{-5}$	$2\cdot10^{-4}$	$3\cdot10^{-4}$	$2\cdot10^{-4}$	$2\cdot10^{-5}$	$2\cdot10^{-5}$	

TABLE 3 (continued)

Metal analyzed (or its compound)	Zn	Cd	In	Sn	Pb	Sb	Bi	Ga	Tl	Te
1	21	22	23	24	25	26	27	28	29	30
Sodium	$2 \cdot 10^{-4}$	$5 \cdot 10^{-6}$ $2 \cdot 10^{-6}$		$2 \cdot 10^{-4}$ $4 \cdot 10^{-5}$	$1 \cdot 10^{-3}$ $4 \cdot 10^{-5}$	$2 \cdot 10^{-4}$	$3 \cdot 10^{-3}$ $6 \cdot 10^{-5}$			
Potassium		$5 \cdot 10^{-6}$		$2 \cdot 10^{-4}$	$1 \cdot 10^{-3}$	$2 \cdot 10^{-4}$	$3 \cdot 10^{-3}$			
Aluminum	$4 \cdot 10^{-4}$	$6 \cdot 10^{-6}$	$3 \cdot 10^{-4}$	$4 \cdot 10^{-5}$	$4 \cdot 10^{-5}$	$4 \cdot 10^{-5}$	$4 \cdot 10^{-5}$			
Calcium	$4 \cdot 10^{-4}$	$6 \cdot 10^{-6}$		$2 \cdot 10^{-5}$	$3 \cdot 10^{-3}$					
Strontium	$2 \cdot 10^{-3}$	$2 \cdot 10^{-5}$	$3 \cdot 10^{-5}$	$5 \cdot 10^{-5}$	$2 \cdot 10^{-3}$	$3 \cdot 10^{-4}$	$1 \cdot 10^{-4}$			
Barium	$2 \cdot 10^{-4}$	$3 \cdot 10^{-6}$		$3 \cdot 10^{-5}$	$2 \cdot 10^{-3}$	$3 \cdot 10^{-4}$	$1 \cdot 10^{-3}$			
Lead	$1 \cdot 10^{-4}$	$3 \cdot 10^{-6}$	$1 \cdot 10^{-5}$	$3 \cdot 10^{-5}$		$3 \cdot 10^{-4}$	$3 \cdot 10^{-5}$		$1 \cdot 10^{-5}$	$5 \cdot 10^{-3}$
Bismuth	$5 \cdot 10^{-4}$	$2 \cdot 10^{-6}$	$5 \cdot 10^{-6}$	$3 \cdot 10^{-4}$	$3 \cdot 10^{-4}$	$5 \cdot 10^{-5}$			$1 \cdot 10^{-3}$	$2 \cdot 10^{-4}$
Gallium	$2 \cdot 10^{-4}$	$2 \cdot 10^{-6}$			$4 \cdot 10^{-5}$		$6 \cdot 10^{-5}$			
Germanium	$2 \cdot 10^{-4}$	$2 \cdot 10^{-6}$	$6 \cdot 10^{-6}$	$2 \cdot 10^{-5}$	$2 \cdot 10^{-5}$	$8 \cdot 10^{-5}$	$6 \cdot 10^{-5}$		$2 \cdot 10^{-5}$	
Tin	$2 \cdot 10^{-3}$	$5 \cdot 10^{-6}$	$1 \cdot 10^{-4}$		$1 \cdot 10^{-4}$	$1 \cdot 10^{-3}$	$1 \cdot 10^{-4}$			
Silicon		$5 \cdot 10^{-5}$		$5 \cdot 10^{-5}$	$2 \cdot 10^{-4}$	$2 \cdot 10^{-4}$	$1 \cdot 10^{-4}$			
Gold	$1 \cdot 10^{-2}$	$1 \cdot 10^{-4}$		$2 \cdot 10^{-3}$	$2 \cdot 10^{-3}$	$3 \cdot 10^{-3}$	$1 \cdot 10^{-3}$			
Magnesium		$1 \cdot 10^{-5}$		$6 \cdot 10^{-5}$	$2 \cdot 10^{-4}$	$1 \cdot 10^{-2}$	$6 \cdot 10^{-5}$			
Molybdenum	$2 \cdot 10^{-3}$	$2 \cdot 10^{-5}$ $1 \cdot 10^{-6}$	$6 \cdot 10^{-4}$	$4 \cdot 10^{-4}$	$4 \cdot 10^{-4}$ $4 \cdot 10^{-6}$	$6 \cdot 10^{-4}$ $1 \cdot 10^{-2}$	$6 \cdot 10^{-4}$ $6 \cdot 10^{-5}$			
Iron	$2 \cdot 10^{-4}$	$2 \cdot 10^{-6}$			$4 \cdot 10^{-5}$		$6 \cdot 10^{-5}$			
Nickel	$3 \cdot 10^{-3}$			$1 \cdot 10^{-3}$	$1 \cdot 10^{-3}$					
Chromium		$3 \cdot 10^{-5}$	$5 \cdot 10^{-5}$		$1 \cdot 10^{-4}$		$1 \cdot 10^{-4}$	$5 \cdot 10^{-5}$		

* The rare earths were determined by the activation method.

B	Si*	As	P,3***	Impurities determined by chemical methods
31	32	33	34	35
$3 \cdot 10^{-4}$			$5 \cdot 10^{-6}$	chlorine $2 \cdot 10^{-4}$; $\qquad 2 \cdot 10^{-4}$ mercury
$3 \cdot 10^{-4}$			$5 \cdot 10^{-6}$	chlorine $2 \cdot 10^{-4}$; $\qquad 2 \cdot 10^{-4}$ mercury
	$3 \cdot 10^{-3}$			
$3 \cdot 10^{-5}$		$2 \cdot 10^{-4}$	$5 \cdot 10^{-7}$	chlorine $5 \cdot 10^{-4}$
$3 \cdot 10^{-5}$		$5 \cdot 10^{-5}$	$5 \cdot 10^{-7}$	chlorine $5 \cdot 10^{-4}$
	$1 \cdot 10^{-3}$			

TABLE 4

Percent Extraction of Impurity Elements into Aqueous Phase (concentrate) during Separation of the Base (bismuth) by Precipitation as BiI_3 with Hydriodic Acid in a Dilute HNO_3 Solution

Impurity elements	Concentrations of impurity elements in artificial mixtures (in % of weight of Bi)					
	$5 \cdot 10^{-1}$	$2,5 \cdot 10^{-1}$	10^{-2}	10^{-3}	10^{-4}	10^{-5}
	% extraction of impurities into concentrate (average results)					

Elements extracted into the aqueous phase (concentrate)

Mg	100	100		90	85	
Ca	99,6	100	93	80	~100	
Ba	99,8	100		90	~100	
Al	99,8	100		93	90	
Ti	98	100		100	~100	
V	99,8	100	100	100	80	85
Cr		100	100	92	90	
Mo	98,1		99	95	96	
Mn	100	100	96,2	90	98	92
Fe	97,7	100	97,7	98	~100	
Co		100	100	98	100	80
Ni	97,4	100	100	97	96	
Zn		100	99	90	~100	
Cd		97,6	90	98	100	~100
In					100	90
Sb		98	96	100	98	~100
Te		95		95	90	
Sm						98,5

Elements precipitated with BiI_3 (not extracted into concentrate)

Ag	0,01			<5	<10	<10
Au			<1		<10	
Pb			<3		< 5	

As an example, Table 4 gives the results of experiments on the determination of the percent extraction of impurity elements into the concentrate (aqueous phase) in the separation of the base (bismuth) by precipitation in the form of BiI_3 with hydriodic acid in dilute nitric acid solution.

Under conditions of mass control, it is most convenient to prepare the impurity concentrates in the form of powders. For this the separated impurities are fixed on a small amount of the corresponding base. Two cases of analysis are possible.

First Case

If the base element has a simple spectrum and satisfies the necessary conditions of spectral analysis, then it is used as the base for the preparation of the impurity concentrate and standards. An unenriched sample is also prepared by direct conversion of the product investigated into the same final form as that in which the chemical concentrate is obtained. During analysis, the spectra of the concentrate and the unenriched sample with single standards are photographed together onto the same plate and under the same conditions. The exposure usually corresponds to complete combustion of the sample, together with the crater of the carbon electrode (2-2.5 min.).

The increase in the relative sensitivity due to the concentration may be illustrated by the two following examples.

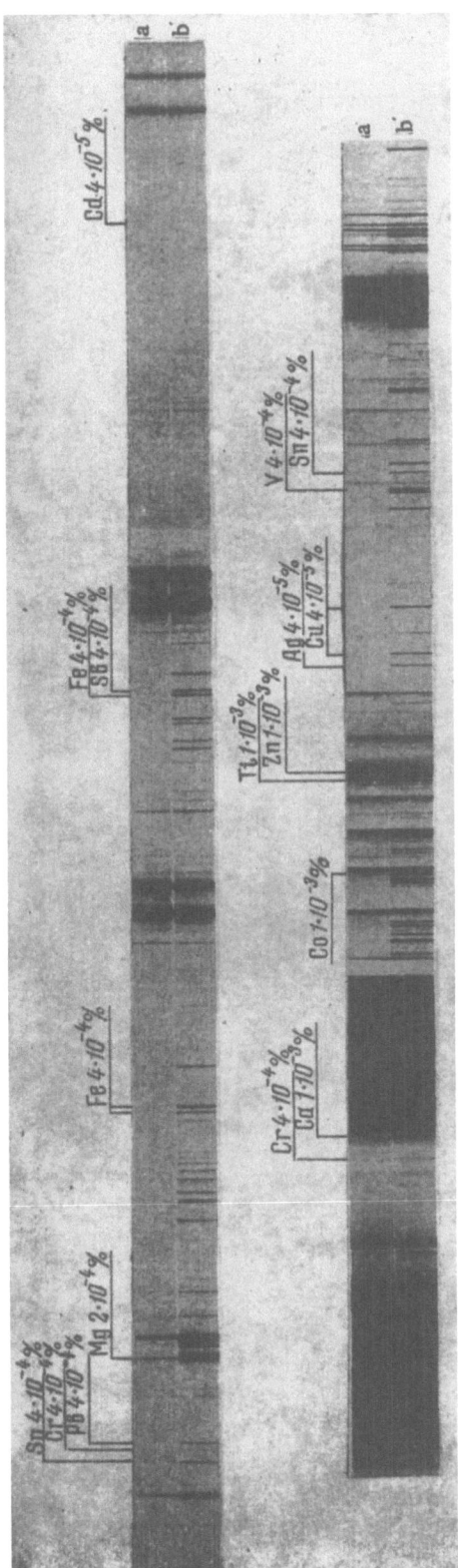

Fig. 1. Comparison of the spectrogram of an aluminum sample (as Al_2O_3) with the spectrogram of the impurity concentrate of the same sample on the same base with an enrichment factor $k = 30$: a) spectrogram of aluminum sample; b) spectrogram of impurity concentrate.

In the analysis of aluminum, the enrichment factor of the impurities reaches ~50, and 23 elements are determined simultaneously.

Figure 1 shows spectrograms of an artificial sample of aluminum oxide without enrichment (upper spectrum a) and the impurity concentrate obtained from the same sample by separating the aluminum chloride (lower spectrum b).

In the first spectrogram (a), the impurity lines are absent or hardly noticeable, while in the second (b), the lines of the same impurities are extremely intense. Therefore, from the spectrum of the unenriched sample it is possible to determine only the impurities present in large amounts (for example, Mg) and those present in small amounts are determined from the spectrum of the concentrate.

In the analysis of germanium, a 6-12 g sample of the metal or the corresponding amount of dioxide yielded, after distillation of the $GeCl_4$ from hydrochloric acid solution, ~170 mg of impurity concentrate in the form of GeO_2 (enrichment factor ~50-100). Twenty-three elements were determined in the concentrate from one spectrogram. The efficiency of concentration is shown in Fig. 2 by comparison of the spectrograms: I — an artificial sample of germanium dioxide without enrichment, and II — a concentrate obtained from the same sample, based on GeO_2.

Second Case

If the base element has a complex spectrum, then two variations are possible.

When the base is of low volatility and it is possible to use the method of fractional distillation with carrier ($AgCl$, Ga_2O_3, etc.) for its spectral analysis, the impurity concentrate, unenriched sample and standards are prepared on this base, mixed with carrier, and the spectra photographed onto one plate. The exposure is chosen on the base of evaporation curves of the impurities and usually is 40-60 sec.

In the second variation, the impurity concentrate is prepared by practically completely separating the base element (to noninterfering concentrations) and an appropriate, spectrally pure compound of a foreign element used as the base for concentrate and standards. The impurity concentrate only is analyzed. Beryllium oxide, in a granular, vitreous form, is very suitable for such a universal base for concentrates. It has a simple spectrum and ensures a steady input of sample into the discharge of a direct current arc. Strontium sulfate was also used as a universal base with satisfactory results.

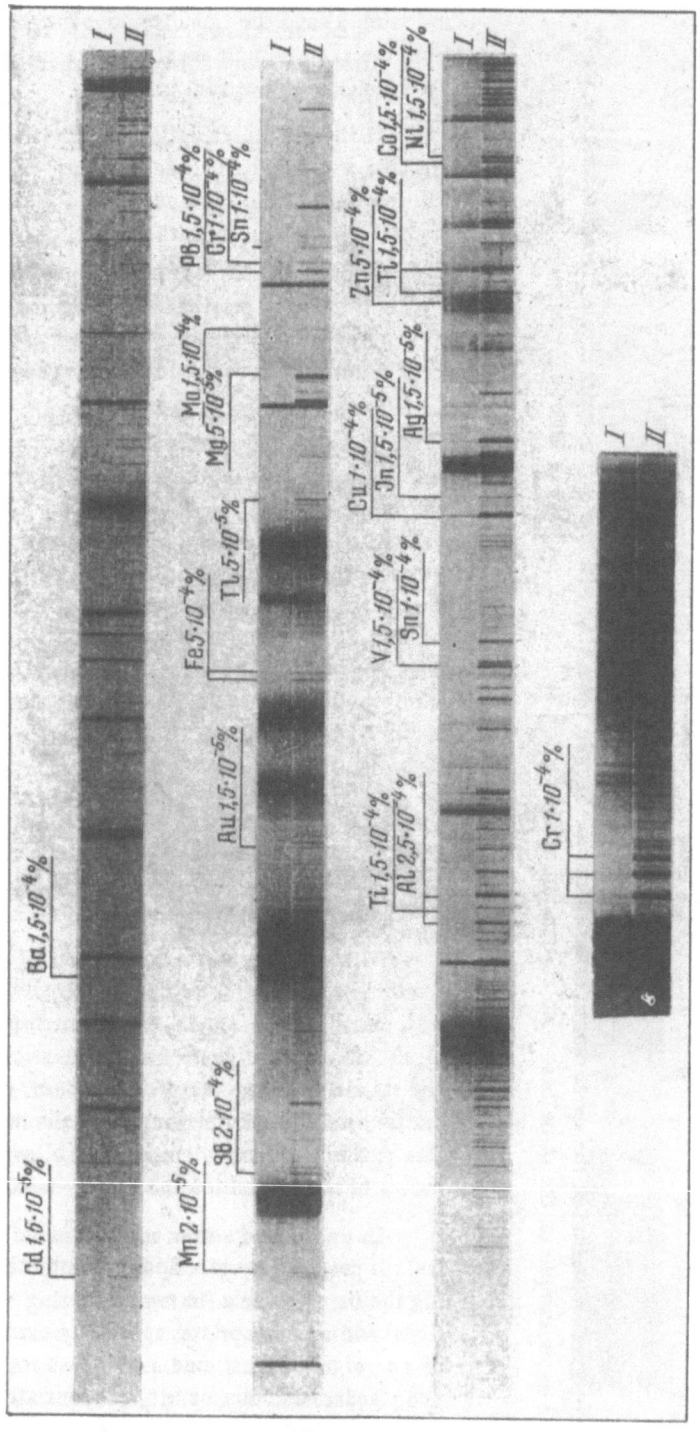

Fig. 2. Comparison of spectrogram I of a germanium dioxide sample (artificial mixture) with spectrogram II of the impurity concentrate obtained from the same sample and based on GeO_2 (enrichment factor $k = 40$).

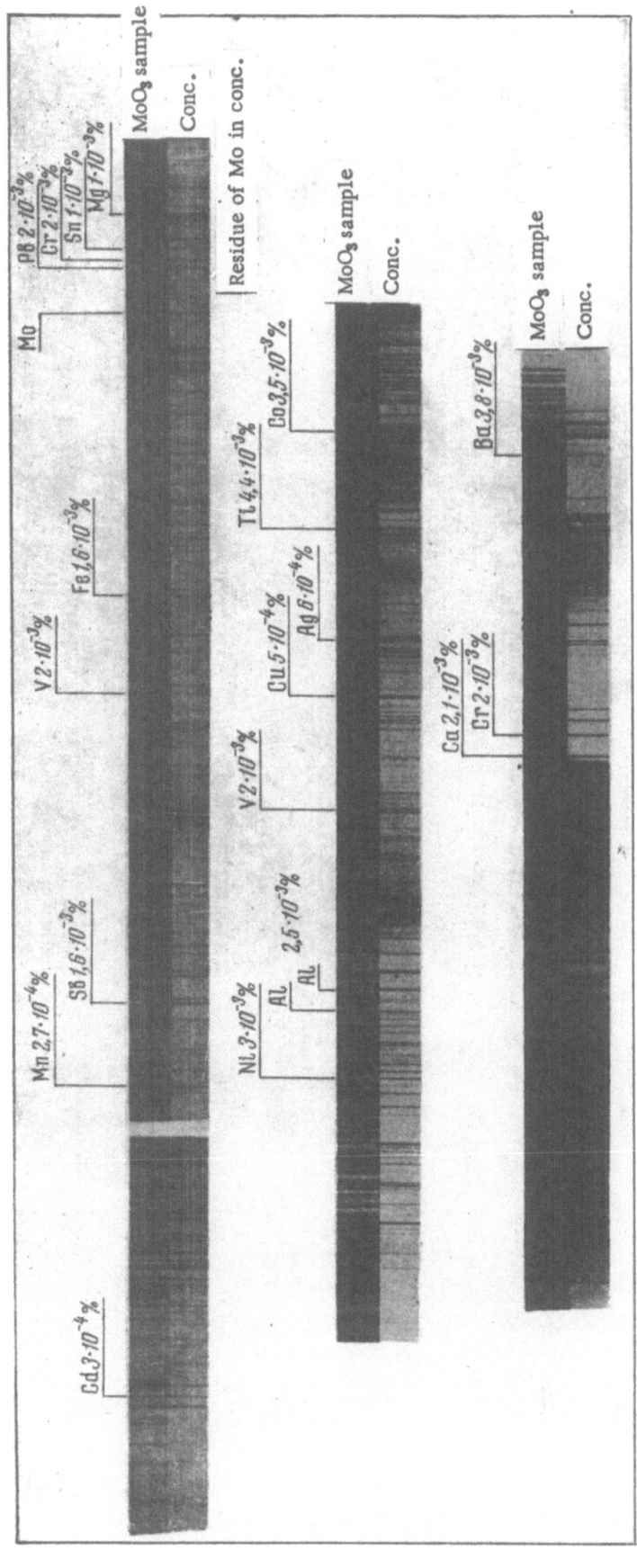

Fig. 3. Comparison of the spectrogram of a molybdenum sample (MoO₃) with the spectrogram of the impurity concentrate of the same sample, based on BeO.

39

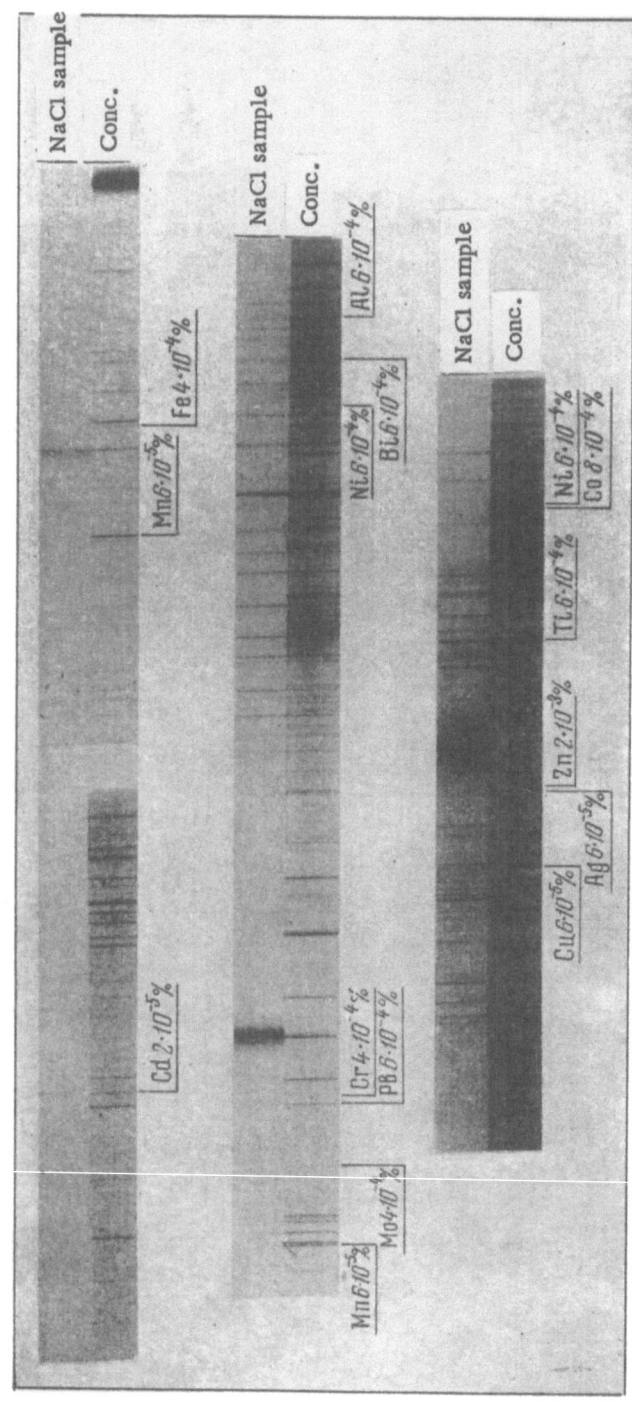

Fig. 4. Comparison of the spectrogram of a sodium sample (as NaCl) with the spectrogram of the impurity concentrate from the same sample, based on BeO. Enrichment factor k = 50.

The use of a universal base for impurity concentrates is also advantageous in those cases where the element investigated is expensive (e.g., gold), radioactive, or particularly objectionable in spectral analysis.

Figure 3 compares the spectrograms of an artificial sample of molybdenum trioxide (upper) and the impurity concentrate obtained from the same sample after separation of the molybdenum by flotation of the compound with α-benzoinoxime (lower spectrogram); the base was beryllium oxide. The impurity lines are extremely clear on the concentrate spectrogram. Due to the complete removal of the molybdenum, whose spectrum contains about 4000 lines, the concentrate could be analyzed on a normal spectrograph of average dispersion and, in addition, an enrichment factor of 10 was achieved. Twenty-one elements were determined in the concentrate simultaneously.

In the analysis of alkali metals and magnesium another enrichment principle was also used — extraction of impurities by group organic reagents. The extraction was carried out both by means of direct extraction with organic solvents and by coprecipitation of traces of elements with organic carriers [37-39]. Hydroxyquinoline was used as the organic carrier.

Figure 4 compares the spectrogram of an artificial sample of sodium (as NaCl) with the spectrogram of the impurity concentrate (based on BeO) obtained from the same sample by extraction with dithizone and hydroxyquinoline.

The standards were prepared as powders by adding to the spectrally pure base all the simultaneously determinable elements and were the same for the analysis of concentrates and unenriched samples.

The spectral analysis conditions for different metals are quite close. In all cases a spectrograph of average dispersion (ISP-22 or ISP-28) with a three-lens condenser was used. The slit width was 10μ. In front of the slit was placed a two-stage attenuator with 100% and ~10% as the relative transmissions of the stages, and due to this it was possible to determine elements, whose analysis lines were in the region 2300-5000 A.

The electrodes were carbon or graphite. The spectrum excitation source was a direct current arc at a current strength of 12 amp and a voltage of 250 v.

The impurities were determined quantitatively by microphotometry, using calibration curves constructed from standards in the coordinates $\log I - \log C$. The blackening measurements were converted into logarithms of intensity, allowing for the blank, on a calculating apparatus designed by G. E. Kriuger.

Tables 1 and 2 give brief characteristics of the spectrochemical methods developed for the analysis of 18 metals: sodium, potassium, aluminum, calcium, strontium, barium, lead, bismuth, gallium, germanium, tin, silicon, gold, magnesium, molybdenum, iron, nickel and chromium. For each of these metals a combined quantitative determination was performed from one spectrogram of a large number (15-25) of impurity elements, usually the following: Be, Mg, Ca, Ba, Al, Ti, V, Cr, Mo, Mn, Fe, Co, Ni, Cu, Ag, Au, Zn, Cd, In, Ge, Sn, Pb, Sb, Bi, Ga, Tl, Te, Pt.

The impurities B, Si and As were determined separately and Li, Na and K, using a glass spectrograph (ISP-51). These impurities include the majority of the metallic elements, which occur in the first five abundance decades [40].

The methods developed are generally applicable; with the use of 4-5 standards, the same for concentrates and unenriched samples, the procedures were applied to the determination of impurities in metals and their compounds over a wide range of concentrations (from $10^{-1}\%$ to $10^{-5}-10^{-6}\%$). In all cases spectral analysis of the sample without impurity enrichment may be used independently to control technical products and to speed the analysis.

Data on the sensitivity of the spectrochemical methods are presented in Table 3.

The accuracy and reproducibility of the methods developed were tested on artificial mixtures, which contained known amounts of trace elements and were subjected to the complete analysis procedure. Parallel chemical and spectral analyses were also carried out and many of them were performed in different laboratories.

The accuracy of the spectrochemical methods, at an element concentration of $10^{-3} - 10^{-5}\%$, was such that the relative error was approximately 20% at a reproducibility of approximately 10%.

Periodic Table

Groups of Elements

Periods	Rows	I	II	III	IV	V	VI	VII (H)	VIII			0
1	I	H 1 — 1.0080										He 2 — 4.003
2	II	Li 3 — 6.940	Be 4 — 9.013	B 5 — 10.82	C 6 — 12.010	N 7 — 14.008	O 8 — 16.0000	F 9 — 19.00				Ne 10 — 20.183
3	III	Na 11 — 22.997	Mg 12 — 24.32	Al 13 — 26.97	Si 14 — 28.06	P 15 — 30.98	S 16 — 32.066	Cl 17 — 35.457				A 18 — 39.944
4	IV	K 19 — 39.096	Ca 20 — 40.08	Sc 21 — 45.10	Ti 22 — 47.90	V 23 — 50.95	Cr 24 — 52.01	Mn 25 — 54.93	Fe 26 — 55.85	Co 27 — 58.94	Ni 28 — 58.69	
4	V	Cu 29 — 63.54	Zn 30 — 65.38	Ga 31 — 69.72	Ge 32 — 72.60	As 33 — 74.91	Se 34 — 78.96	Br 35 — 79.916				Kr 36 — 83.7
5	VI	Rb 37 — 85.48	Sr 38 — 87.63	Y 39 — 88.92	Zr 40 — 91.22	Nb 41 — 92.91	Mo 42 — 95.95	Tc 43 — [99]	Ru 44 — 101.7	Rh 45 — 102.91	Pd 46 — 106.7	
5	VII	Ag 47 — 107.880	Cd 48 — 112.41	In 49 — 114.76	Sn 50 — 118.70	Sb 51 — 121.76	Te 52 — 127.61	I 53 — 126.92				Xe 54 — 131.3
6	VIII	Cs 55 — 132.91	Ba 56 — 137.36	La 57 ★ — 138.92	Hf 72 — 178.6	Ta 73 — 180.88	W 74 — 183.92	Re 75 — 186.31	Os 76 — 190.2	Ir 77 — 193.1	Pt 78 — 195.23	
6	IX	Au 79 — 197.2	Hg 80 — 200.61	Tl 81 — 204.39	Pb 82 — 207.21	Bi 83 — 209.00	Po 84 — 210	At 85 — [210]				Rn 86 — 222
7	X	Fr 87 — [223]	Ra 88 — 226.05	Ac 89 ★★ — 227	(Th)	(Pa)	(U)					

★ Lanthanides 58–71

58 Ce — 140.13	59 Pr — 140.92	60 Nd — 144.27	61 Pm — [147]	62 Sm — 150.43	63 Eu — 152.0	64 Gd — 156.9	65 Tb — 159.2	66 Dy — 162.46	67 Ho — 164.94	68 Er — 167.2	69 Tm — 169.4	70 Yb — 173.04	71 Lu — 174.99

★★ Actinides

90 Th — 232.12	91 Pa — 231	92 U — 238.07	93 Np — [237]	94 Pu — [239]	95 Am — [241]	96 Cm — [242]	97 Bk — [243]	98 Cf — [244]

Legend:
S 16 — 32,066
— Atomic number
— Chemical symbol
— Atomic weight
— Electron layers

* Circles mark elements, traces of which are concentrated in the acid solution in the separation of molybdenum with α-benzoinoxime.

Besides the methods described, which are designed for the spectral analysis of powders, in a series of spectrochemical methods the impurity concentrates were analyzed as solutions. As is known, the analysis of solutions with spark-excitation of the spectra is distinguished by a high absolute sensitivity [41-43]. In this case, the impurity concentrate, after complete separation of the base element, was obtained as a solution with a minimal volume of ~0.25 ml.

The investigated and standard solutions were applied with micropipettes to the planar ends of graphite electrodes, previously treated with a solution of polystyrene in benzene, and dried [44]. The spectra were excited with a spark generator IG-2 and photographed on an ISP-22 spectrograph. With these methods a high relative sensitivity ($10^{-4} - 10^{-6}\%$) was attained in the analysis of small samples (up to 100 mg).

The methods developed were used for control of industrial production of metals and their compounds of high purity and there is a considerable amount of statistical material on some of them. Practice has confirmed the efficiency of organizing the analytical control of high-purity metals by a combination of two methods, with the close association of spectroscopists and chemists.

LITERATURE CITED

[1] G. Boid, Progr. Phys. Sci. 40, No. 3, 445 (1950).

[2] I. P. Alimarin, Iu. V. Iakovlev and A. I. Zhabin, Collection: The Application of Tracers in Analytical Chemistry (Izd. AN SSSR, 1955) pp. 58-69.

[3] A. C. Wahl and N. A. Bonner, Radioactivity Applied to Chemistry (1951) pp. 88-93, 381-383.

[4] D. Hughes, Neutron Investigations in Nuclear Reactors [Russian translation] (IL, 1954) pp. 237-247.

[5] A. A. Smales and B. D. Pate, Anal. Chem. 24, No. 4, 717 (1952).

[6] V. Meinke, Prog. Chem. 25, No. 6, 770-780 (1956).

[7] A. P. Vinogradov, Collection: Investigations in the Field of Geology, Chemistry and Metallurgy. Reports of the Soviet Delegation to the International Conference on the Peaceful Uses of Atomic Energy (Izd. AN SSSR, 1955) pp. 72-89.

[8] C. J. Rodden (Editor), Analytical Chemistry of the Manhattan Project (1950).

[9] E. B. Sandell, Colorimetric Determination of Traces of Metals [Russian translation] (Goskhimizdat, 1949).

[10] E. B. Sandell, Colorimetric Determination of Traces of Metals (1950).

[11] V. K. Prokof'ev, Photographic Methods of Quantitative Spectral Analysis of Metals and Alloys, Parts I and II (GITTL, 1951).

[12] A. K. Rusanov, Spectral Analysis of Ores and Minerals (Gosgeolizdat, 1948).

[13] F. Twyman, Metal Spectroscopy (1951).

[14] L. H. Ahrens and G. R. Harrison, Spectrochemical Analysis (1950).

[15] N. H. Nachtrieb, Principles and Practice of Spectrochemical Analysis (1950).

[16] Methods for Emission Spectrochemical Analysis (ASTM, 1953).

[17] S. L. Mandel'shtam, N. N. Semenov and Z. M. Turovtseva, J. Anal. Chem. 11, No. 1, 21-29 (1956).

[18] A. N. Zaidel', N. I. Kaliteevskii, L. V. Lipis, M. P. Chaika and Iu. I. Beliaev, J. Anal. Chem. 11, No. 1, 9-20 (1956).

[19] A. N. Zaidel', N. I. Kaliteevskii, A. A. Lipovskii, A. N. Razumovskii and P. P. Iakimova, Bull. Leningrad Univ., No. 22, Div. of Phys. and Chem. No. 4, 18-40 (1956).

[20] A. O. Voinar, Combined Application of Spectrographic, Microchemical, Electrophoretic and Counting Procedures to the Determination of Low Concentrations of Microelements in Biological Objects, Report to the 10th All-Union Conference on Spectroscopy, 1956.

[21] Analytical Chemistry of the Manhattan Project, Div. VIII, vol. I, 339, 396, 410-412, 428, 459-460, 465, 469, 474, 483, 484, 489-491, 640, (1950).

[22] N. H. Nachtrieb, Principles and Practice of Spectrochemical Analysis (1950) pp. 300-314.

[23] L. H. Ahrens and G. R. Harrison, Spectrochemical Analysis (1950) pp. 104, 160-161, 195, 200, 202, 204, 208-210, 212, 216.

[24] F. Twyman, Metal Spectroscopy (1951) pp. 246, 388-391, 440-443, 449-451, 456.

[25] E. B. Sandell, Colorimetric Determination of Traces of Metals (1950) pp. 7, 12.

[26] R. L. Mitchell, Mikrochemie, Vol. 36/37, No. 2, 1042-1047 (1951).

[27] R. L. Mitchell and R. O. Scott, Spectrochim. Acta 3, 367 (1948); J. Soc. Chem. Ind. 66, 330 (1947).

[28] G. E. Heggen and L. W. Strock, Anal. Chem. 25, 6, 859-863 (1953).

[29] F. Burriel-Marti and J. Ramirez-Munoz, Mikrochemie, Vol. 36/37, No. 1, 495-512 (1951).

[30] G. Sempels, Spectrochim. Acta. 3, 246 (1948).

[31] G. Gorbach and F. Pohl, Angew. Chemie 9, 242 (1953).

[32] F. Pohl, Spectrochim. Acta 6, 1, 19-22 (1953).

[33] G. Gorbach and F. Pohl, Mikrochemie, Vol. 36/37, No. 1, 487-494 (1951); Vol. 38, No. 3, 258-267 (1951); Vol. 38, No. 4, 328-341 (1951).

[34] H. Speker and H. Hartkampf, Angew. Chemie 67, 6, 173 (1955).

[35] A. Hulanicki, Wiadom. Chem. 9, No. 5, 249-264 (1955).

[36] A. G. Karabash, Trans. Commission on Analytical Chemistry, Vol. V (VIII) 270-293 (1954).

[37] V. I. Kuznetsov, Factory Labs. 2, 263 (1945).

[38] V. I. Kuznetsov, J. Anal. Chem. 9, 4, 199 (1954).

[39] V. I. Kuznetsov and G. V. Miasoedova, Collection: The Application of Tracers in Analytical Chemistry (Izd. AN SSSR, 1955) pp. 24-28.

[40] A. A. Saukov, Geochemistry (Geolizdat, 1950) pp 50-54.

[41] M. Fred, N. H. Nachtrieb and F. S. Tomkins, J. Opt. Soc. Amer. 37, 4 (1947).

[42] N. H. Nachtrieb, Principles and Practice of Spectrochemical Analysis (1950).

[43] C. J. Rodden (Editor), Analytical Chemistry of the Manhattan Project (1950) pp. 615-619.

[44] Kh. I. Zil'bershtein, Factory Labs. 19, 4 (1953).

QUANTITATIVE SPECTRAL DETERMINATION OF IMPURITIES
IN RADIOACTIVE PREPARATIONS

B. V. L'vov and G. I. Kibisov

In a series of cases it is necessary to know the concentrations of certain impurity elements in standard radioactive preparations. The permissible amounts of the impurities in these preparations are determined by technical conditions.

The determination of very small concentrations of various elements in radioactive preparations by a chemical method is very time consuming and is considerably complicated by the harmful action of radiation. Therefore, in many cases of purity control it is advantageous to use emission spectral analysis, in which small amounts of the radioactive preparations are consumed in determination of the impurities, and the analysis time is considerably shortened.

Due to the characteristics of the objects analyzed, the procedures developed were based on the following principles.

1. The procedures must be of maximum simplicity and be identical for different substances, if possible. For simplification, some analysis stages were excluded, for example, the introduction of an internal standard.

2. The amounts of material required for the analysis must be minimal and the duration of the analysis must be as short as possible. This forced us to use the most sensitive methods of spectral analysis. For the same reason the spectra were excited in an arc discharge and not in a spark.

3. Despite the simplifications, which lower the accuracy of the determination, the analyses must be fully quantitative in character.

4. The analyst must be reliably shielded from aerosols of radioactive substances, formed during the arc discharge and if only partially, from β- and γ-radiation.

5. The shielding apparatus must be as simple as possible.

Shielding Apparatus

In addition to the observation of normal precautions, in the spectral analysis of radioactive substances particular attention has to be paid to shielding personnel from ingestion of aerosols of the radioactive substances into the lungs, since there is always intense atomization of the substance in spectral analysis. For this purpose it is necessary to isolate the arc in which the spectra are excited. There are few data on the construction of such apparatuses in the literature [1-3]. The apparatuses described are designed either for the use of spark discharges only or for the analysis of highly active α-preparations of the transuranium series; moreover, they are extremely expensive and complicated to use and to make.

We had the problem of developing a sufficiently simple apparatus, designed for use on standard spectral equipment and suitable for the analysis of various β- and γ-active preparations. Since the exposures we adopted were small, and the amounts of preparation consumed in a single exposure did not exceed 1 μC (for preparations of average specific activity), in the construction of a chamber for exciting spectra, most attention was not paid to radiation shielding, but to efficient sealing in the inner space to prevent an aerosol escaping into the air. The chamber had to have the minimum possible internal surface, directly contaminated by the atomized material, and efficient cooling.

Fig. 1. General view of chamber with cell tube lowered (operating position).

Fig. 2. Cylinder.

The general appearance of the chamber constructed is given in Fig. 1.

The base of the chamber was a hollow steel cylinder, closed at the ends with silica and glass windows (Fig. 2). In the sides the cylinder had two openings to accommodate the electrode holders and was placed in a brass box, which served as a cooler. The cooler was mounted on a base, which was bolted to the rails of the spectrograph. The steel electrode holders (Fig. 3) were isolated from the body of the chamber with porcelain sleeves. Into the porcelain sleeves were cemented (BF-2 cement) threaded metal sleeves for regulating the position of the holders. A thread was cut in the outer surface of porcelain insulators for a steel tube, which was screwed onto them after assembly of the electrodes and holders. The closed electrodes (cell) were placed in the openings of the chamber, after which the tube was screwed together and lowered to the stop and the ends of the electrodes, forming the arc gap, were approximately on the optical axis. After adjusting the position of the electrodes (by screwing up the steel holders), we connected the generator leads to them and struck the arc discharge. At the end of the exposure, the tube was again screwed to the holders and the cell transferred to a box for cleaning. Another cell was inserted in its place, etc.

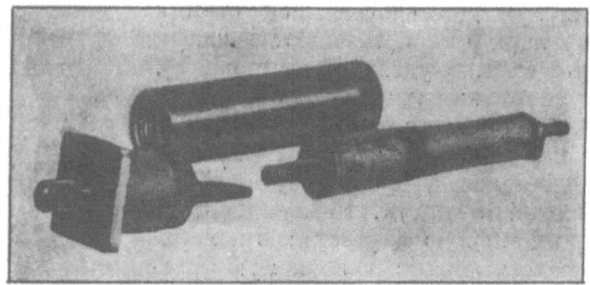

Fig. 3. Dismantled cell with carbon electrodes mounted in holders.

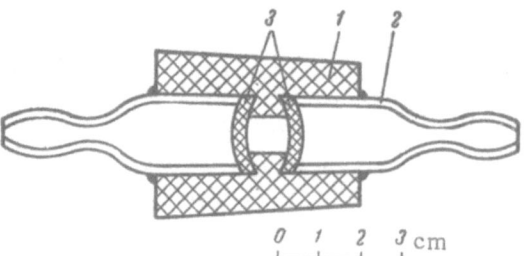

Fig. 4. Cross section of filter cell: 1) bung; 2) connecting tube; 3) filters.

So as to reduce the contamination of the internal surface of the cylindrical cavity of the chamber with active materials, the cavity was sheathed with a polythene film, which was destroyed after the analysis. The film was cut accurately to the dimensions of the cavity and was held tightly inside by its own springiness. For this purpose we could use films of not less than 200μ thick. Thinner films were less springy, came loose from the cooled cylinder surface, buckled and burned.

The atomized substance and the gases were sucked off from the chamber cavity through a double filter with a pump. Due to evacuation of the inner chamber and the passage of air into it from the room through the chinks, there was little probability of active materials escaping into the working space. The filters were fitted into a rubber bung with a shaped hole passing through (Fig. 4). They were pressed tightly against an inner step in the bung opening with glass connecting tubes. After the analysis, the filter and the piece of contaminated tubing connecting the chamber to the filter were destroyed.

The preparation of samples for analysis, mounting of electrodes in the holders, and assembly of the cell were performed in a specially made box. The box was assembled from steel sheets, 5 mm thick, covered with lead on the inside (lead thickness 5 mm). A lead glass observation window was set at an angle in the front wall and rubber gloves were fitted. Inside the chamber was fitted a sink for washing contaminated apparatus and taps for hot and cold water. The inner section of the box was reached through an accessory antechamber.

Analytical Procedure

The manipulations were simplified to some extent by working with solutions. Therefore, in all possible cases, powders were converted into salt solutions. For example, metallic copper powder was converted into $CuSO_4$ solution and $SrCO_3$ into $Sr(NO_3)_2$ solution. Sometimes solution was difficult or involved the use of insufficiently pure reagents. For example, Cr_2O_3 is usually dissolved by fusion with alkalis. This procedure is difficult, causes the introduction of impurities and is also unacceptable because alkali elements are determined in Cr_2O_3. Therefore, such substances were analyzed in the original state.

In the analysis of solutions, samples were introduced into the analytical gap by the method recommended by Zil'bershtein [4]. In work with powders, samples were deposited on carbon electrodes in the form of suspensions in glycerol [5]. In either case, not more than 2-3 mg of preparation was used for a single exposure.

We developed a large number of procedures for spectral determinations of various elements in various preparations. As examples, we will describe two procedures, the determination of P and Fe in CdS powder, and the determination of Co , Mn and Sb in $FeCl_3$ solutions. Standards were prepared from normal inactive materials by the introduction of definite volumes of solutions into the purified base. If it was impossible to remove the elements determined from the base completely, then the amount of the residual impurity was evaluated by the method of additions. A 100 mg sample of CdS powder was mixed for 3 min with 1 ml of glycerol and one drop of the suspension was deposited on the end of a carbon electrode, ground flat. The carbons were preliminarily burnt off under draft in a D.C. arc of 30 amp for 15 sec. The electrodes on the metal bases were dried off on a hot-plate until the glycerol evaporated completely. The spectra were excited in a D.C. arc at 6 amp for 15 sec. The analytical gap was 1.5 mm. An ISP-22 spectrograph with three-lens illumination of the slit was used. The slit width was 18μ. The plates used were spectral, type 2, with a sensitivity of 16 All-Union State Standard units. The development was for 6 min. in developer D-11 at 20°. The range of concentrations determined (in % of CdS) and the analytical pairs of lines were: 0.003-0.125% – Fe 3020.6 and Cd 2733.9 A; 0.001-0.01% – P 2535.6 and Cd 2733.9 A. Calibration curves were constructed for each plate from four standards in the coordinates $\log I_{an}/I_{Fe}$, $\log C$. The graphs were straight lines and had a slope of approximately 1. There was a slight, parallel displacement of the graphs. The probable error, characterizing the reproducibility with three exposures of the spectra, was 7% for Fe and 6% for P.

In the analysis of $FeCl_3$ solution (with an Fe concentration of ~50 mg/ml) for the Co, Mn and Sb contents, the lower carbon electrodes, which were ground flat, were treated with a 3% solution of polystyrene in benzene by immersing the ends of the electrodes in the solution for several seconds. When the electrode was dry, it was treated with one drop of 5% sulfuric acid and heated on a hot-plate until the H_2SO_4 had decomposed completely. Then one drop of the $FeCl_3$ solution was deposited on the cold electrode, which was dried with gentle heating. This procedure enabled the solution to penetrate the depth of a thin surface layer of carbon, as no visible residue remained on the end. The spectra were excited and recorded on the same apparatus, with the same parameters as in the previous procedure. The analytical pairs of lines and the concentration range of impurities in percents of Fe were as follows: 0.05-0.3% – Sb 2311.5 and Fe 2320.4 A; 0.005-0.3% – Co 2424.9% and Fe 2390.0 A; 0.005-0.3% – Mn 2933.1 and Fe 2390.0 A. Calibration graphs, plotted in logarithmic coordinates, were straight lines with slightly different slopes for different plates: from 0.67 to 1.0 for Sb, from 0.74 to 0.8 for Co, and from 0.9 to 1.0 for Mn. The probable error of the analysis with three exposures of the spectra was 7% for Sb, 6% for Co, and 3.5% for Mn.

A calculator was used to speed the analyses. With the drying of the plates accelerated and the standard spectra exposed beforehand, the results of an analysis may be obtained within an hour of receiving the sample.

CONCLUSIONS

Considering the difficulties and the limitations, arising in the spectral analysis of radioactive preparations, the procedures developed may be considered quite satisfactory for purity control of active preparations.

However, it is obvious that the results obtained are not in any way final, and that further work is to be done both in developing the shielding apparatus and in improving the separate stages of the analysis.

LITERATURE CITED

[1] C. Feldman and M. B. Hawkins, Anal. Chem. 22, 1400 (1950).

[2] J. C. Conway and M. F. Moore, Anal. Chem. 24, 463 (1952)

[3] F. T. Birks, Spectrochim. Acta 8, 167 (1956).

[4] Kh. I. Zil'bershtein, J. Tech. Phys. (USSR) 25, 8 (1955).

[5] R. C. Hughes, Anal. Chem. 24, 1406 (1952).

THE PRODUCTION OF ALPHA-, BETA- AND GAMMA-SOURCES
USING OXIDE FILMS ON ALUMINUM AND ITS ALLOYS

M. S. Petrova

In a series of fields in medicine, science and technology, radioactive isotopes are used as radiation sources. Sources with various dimensions and configurations and different activities and radiation characteristics are required.

The problem of the present work was to determine the possibilities of using for the preparation of low-activity sources the properties of oxide films on aluminum and its alloys, filled with various chemical reagents [1]. The properties of the oxide film and its filling capacity were studied in detail by G. V. Akimov et al. A solution of the problem would make it possible to prepare sources on flexible foils and on rods, needles, balls and components with other forms, with a practically unlimited choice of isotope.

It was established by x-radiography and electron diffraction that the films obtained anodically on aluminum were mainly amorphous aluminum oxide; the films must be almost nonporous (for example, for rectifiers), thin, or comparatively thick, porous, coarse and elastic.

The character and properties of the films depend on the conditions of their formation, the composition of the electrolyte, the current characteristics (voltage, density and A.C. or D.C. current), temperature, exposure time, and many other factors; the material and its structure, and also the mechanical treatment of the surface, are of great importance in the growth of the film [2, 3].

Under standard conditions, used in the sulfuric acid anticorrosion anodic oxidation of aluminum, films are obtained with a thickness of up to 10-15 μ and a pore diameter of 0.1 μ, but by changing the conditions, it is possible to increase the film thickness to 200 μ. Edwards and Baumann, and also other authors [4,5] found that the porosity of the films obtained under these conditions reached 36% of the total film volume and that under definite conditions the pores of freshly oxidized films were rapidly filled with various chemical reagents. Similar data on the porosity of anode films on aluminum were also described in the latest work of A. F. Bogoiavlenskii [5].

From the work of N. D. Tomashov and M. N. Tiukina [1], however, it follows that the filling of the film cannot be considered as a simple "impregnation." There may be considerable changes in the physical and physicochemical characteristics of the film, depending on the filling conditions. Thus, for example, as practice has shown [6], films obtained by oxidation in oxalic acid with A.C., have a considerable thickness and porosity and are capable of being filled strongly with acid dyes, while for high-grade filling with basic dyes, a preliminary additional treatment of the film is desirable.

A filled film is stable to washing in water, mechanical treatments, and other tests to the same degree as an unfilled film. The corrosion stability is sharply increased by anticorrosion fillers. The hardness of the film on aluminum varies from 300 to 1500 kg/ cm^2, depending on the purity of the aluminum. If we remember that tempered steel, U-10, has a hardness of approximately 1000 and cast aluminum, 50 kg/ cm^2, it is obvious that the hardness given for the oxidized aluminum is very high. According to the data of N. D. Tomashov, the wear of oxidized aluminum paired with steel falls to a loss of 2.5 mg compared with 15 mg under the same conditions without an oxide layer.

In the experiments on the filling of films with solutions containing radioactive isotopes considered below, we used the standard sulfuric acid method of anodic oxide formation, which gave a sufficiently high film porosity.

The experiments described below may be considered as a search to try out the possibilities of preparing solid samples by a general technological process without reference to any particular case, as direct indications on this problem are absent from the literature. The works of A. F. Bogoiavlenskii et al. consider the introduction of isotopes into oxide films directly during the electrolysis at the moment of oxide formation by the introduction into the electrolyte of various compounds related to it in structure and containing isotopes, i.e., in this case there is a different technological process, whose us is considerably limited in range and activity limits, and involves a complicated apparatus.

Together with the principle problem of the possibility of preparing β- and γ-sources through oxides, in passing, an examination was made of the effect of oxidation time, concentration, and specific activity of "filler" solutions, their acidity, and the filling time.

Experimental Conditions

Plates of A1 grade aluminum, 0.8 mm thick, and measuring 40 × 40 mm, were used for the experiments. All the plates were stamped out.

Oxidation conditions. Sulfuric acid electrolyte containing 200 g/liter of sulfuric acid, $D_{an} = 0.015$ amp/cm^2, V = 16 v, t = 18°; before the oxidation, a standard preliminary preparation was performed: mounting on contact arrangement, degreasing, cleaning, and the appropriate washings.

The filling of the films was performed directly after the anodic oxidation and washing (not more than a 1-2 hour interval), at room temperature, by immersion in the filler solution containing the radioactive isotope. Approximately 1 ml of solution had to be used per cm^2 of oxidized surface.

After being kept in the fillers for the given time, the filled films were thoroughly washed with warm running water while the surface was rubbed with a soft gauze pad, dried, and tested for activity. The measurements were performed on apparatus "B" under completely identical conditions.

Experimental Data

Table 1 gives the characteristics of the solutions used for filling. The choice of isotopes for the filling solutions was conditioned first by the demand for sources with that type of radiation (Co^{60}, Tl^{204}, P^{32} and C^{14}), and second by the presence of the appropriate solutions; less concentrated solutions were prepared by diluting part of the original solutions.

TABLE 1

Solution No.	Isotope	Compound in which the isotope was present	Salt content mg/ml	Acidity	Specific activity μC/ml
1	Co^{60}	Co(NO$_3$)$_2$	0,008	0,01 N	0,70
2	Co^{60}	the same	0,004	0,005 N	0,36
3	Zn^{65}	ZnSO$_4$	0,16	pH = 6,2	0,002
4	Zn^{65}	the same	0,08	pH = 6,2	0,001
5	Tl^{204}	TlSO$_4$	0,02	pH = 6,8	0,0025
6	P^{32}	NaH$_2$PO$_4$	2,0	pH = 6,8	0,0050
7	P^{32}	the same	1,0	pH = 6,8	0,0025
8	C^{14}	NaHCO$_3$	0,05	pH = 7,1	0,0023

All the filling experiments were performed with three groups of oxidized plates, differing in time of anodic treatment. Three anodic oxidation times, 5, 15 and 30 min., and two filling times, 2 and 10 min., were used.

Table 2 gives the final data of all the primary experiments; the activities of the samples are given in counts/min. as the arithmetic mean of two parallel determinations; we considered only the data which remained constant after several washings and had a "scatter" of less than 20%.

TABLE 2

Isotope	Solution No.	With a filling time of 2 min			With a filling time of 10 min		
		oxidation exposure			oxidation exposure		
		5 min	15 min	30 min	5 min	15 min	30 min
Co[60]	1	Entire series of exp't. gave a "scatter of more than 50%					
	2	160	160	176	256	320	384
Zn[65]	3	272	384	412	528	720	816
	4	144	224	288	240	416	464
Tl[204]	5	384	—	1760	784	—	2800
P[32]	6	2480	8000	15360	3284	10800	17920
	7	2080	7800	13120	2816	8800	16640
C[14]	8	288	1104	2016	608	3008	4304

In addition to the experiments described above, several experiments were performed with solutions of Co[60], Tl[204] and P[32] with sixteen hours treatment with the filler. The data obtained confirmed the results of the short duration "filling," namely, that the samples from Co[60] solutions had a large "scatter" and a low activity. The activity of samples with Tl[204] and P[32] had a considerable scatter and reached the order of $10 \mu C/cm^2$.

Simultaneously with the experiments described above, twelve experiments were performed on film filling in solutions of sodium sulfate with the isotope S[35]. The filling solutions used had various concentrations, but the pH was kept at 7.0-7.1. Here, plates from different oxidation conditions (as regards exposure) were used; they were filled for from 2 min. to 16 hours. In all the cases, we could not observe filling of the film, registerable on the counter, after washing the samples.

The experimental data on the filling of electrochemically prepared aluminum oxide films by a tracer method showed that the exposure time in the range from 5 to 30 min. increased the capacity of the films for filling, but there was not a linear dependence.

From the experiments with Zn[65] and phosphorus, it follows that an increase in the activity and the concentration intensifies the filling. However, in the filling of films with phosphate solution, the process proceeds with such intensity that the activity increases extremely slightly under these conditions with an increase in the solution concentration for identical plates and a filling time of from 2 to 10 minutes. Changing the oxidation time over the limits tested made it possible to change the activity by a factor of more than 10.

Considering that the anodic oxidation was performed in a sulfuric acid bath, one might expect that there would be no fixing of radioactivity in samples filled in solutions of sodium sulfate with S[35], since the active parts of the aluminum oxide would already be filled with SO_4 ions during the electrolytic formation of the film and additional filling of the film with SO_4 ions with S[35] would only be possible due to isotope exchange, which is obviously of no practical importance under these conditions.

Experiments with Co[60] showed that the films could not be filled in even weakly acid solutions as the oxide dissolved.

A substantial addition to the preparation of fixed films from various types of solutions with radioactive isotopes is the application of fillers making use of chemisorption and the formation of insoluble compounds, as is recommended for dyes [7, 8], for example, and this must be used in future work, particularly in those cases where difficulties arise in the direct filling of a film.

As has already been stated, the films obtained were mechanically strong and did not produce active contamination when other objects came into direct contact with their surfaces and also did not change their activities under prolonged and intensive washing with water.

CONCLUSIONS

1. The filling of electrochemically prepared aluminum oxide film shows its typical behavior as an adsorbent to which may be extended all the rules pertaining to aluminum oxide as a chromatographic adsorbent [9].

2. We can recommend the method of "filling" freshly prepared (electrochemically) aluminum oxides with various isotopes as a general procedure for preparing α- *, β- and γ-sources of low activity, differing in form and dimensions and in activity and characteristics of the radiation.

The method makes it possible to obtain local radiation on components by isolating the appropriate place during oxidation and also to obtain portions with different radiation characteristics on the same component by successively isolating and filling the separate portions.

By changing the oxidation conditions and the specific activity and other characteristics of the filler solutions, it is possible to obtain sources with a wide range of activities without inefficient use of the isotope.

3. Practically without arranging complicated shielding equipment and manipulators, it is possible to prepare technical sources from aluminum rod in the form of cylinders, needles, applicators, etc., e.g., with Co^{60}, and large plates for dispersing electrostatic charges in the textile and printing industries, for example, with thallium or other isotopes. It would also be possible to prepare many other articles with a standard anodic oxidation bath with radiation shielding used only for the filling (dipping) and washing sections, i.e., on simple technological operations.

4. The organization of a source production line using the procedure proposed would require for each isotope and activity range the development of a technological process of preparing and measuring them and, at the same time, by using tracers, it is possible to study thoroughly the formation conditions and properties of oxide films with different anodic oxidation conditions and their dependence on the structure of the metal.

LITERATURE CITED

[1] G. V. Akimov (Editor), Collection: Rapid Methods of Protecting Components from Corrosion (Izd. AN SSSR, 1946); G. V. Akimov, N. D. Tomashov and M. N. Tiukhina, An investigation of the anodic treatment process, p. 7; G. V. Akimov, N. D. Tomashov and M. N. Tiukhina, Development of rapid oxidation conditions, p. 25; N. D. Tomashov and M. N. Tiukhina, Development of a rapid method of color filling of oxide films, p. 45; N. D. Tomashov and M. N. Tiukhina, Mechanism of color filling of anodic oxide films on aluminum, p. 57.

[2] G. S. Vozdvizhenskii, A. Sh. Valeev and T. N. Grechukhina, Proc. Acad. Sci. 72, No. 2, 311 (1950). See also J. Phys. Chem. 25, 1, 87 (1951).

[3] J. D. Edwards and F. Keller, Metals Technology 1700 (April, 1944).

[4] W. Z. Baumann, Z. Physik 3, 708 (1939).

[5] A. F. Bogoiavlenskii and N. B. Siletskaia, J. Appl. Chem. 29, 8, 1925 (1956).

[6] Petrova, A.C. Anodic Oxidation of Aluminum Alloys, TEKSO No. 350/16 (1943).

[7] G. M. Badal'ian, New Process for Chemical Coloring of Oxidized Surfaces of Aluminum Alloys, TEKSO No. 1209/19 (1950).

[8] V. D. Teitel'man, Precision Industry 10, 20 (1940).

[9] K. M. Ol'shanova and K. V. Chmutov, J. Anal. Chem. 8, 211 (1953).

[10] V. O. Krening and R. S. Ambartsumian, Corrosion of Metals in Aviation (Oborongiz, Moscow:1941).

[11] L. I. Kadaner, Protective Films on Metals (Izd. Kharkov. Univ., 1956);

---

* A solution of polonium, for example, could be used as the filler for preparing α-sources.

* * Card index for the exchange of production experience.

STABLE ISOTOPES ENRICHED BY THE ELECTROMAGNETIC METHOD

V. S. Zolotarev

The development of isotope separation methods over the last fifteen years has made it practical and possible to prepare enriched isotopes of any element.

Of the various methods of separating isotopes, the electromagnetic method is the main one, as it combines particularly favorable specific characteristics, general applicability, making it possible to separate isotopes of any element, exceptionally high separation efficiency in one stage, the possibility of separating all the isotopes of the given element simultaneously, and maximum flexibility allowing a rapid change from the separation of the isotopes of one element to the separation of those of another.

Thanks to this, the electromagnetic method became the main one for satisfying the demand for a wide range of isotopes of various elements. The isotopes of 44 elements have been separated in milligram and gram amounts on USSR apparatuses.

We are not concerned with the general principles and the technological details of the method. The first are quite well known and a detailed description of the method is impossible within the framework of the present report. Therefore, we are limited to information only on the main characteristics of the electromagnetic method, which are of interest to a wide range of investigators working or planning to work with enriched isotopes.

Capacity

The capacity of the electromagnetic method is determined in the first place by the value of the ion current obtained from the ion source. The value of the ion current is not constant and varies from element to element, from tens to hundreds of milliamperes, depending on the mass of the isotopes separated, the relative content of neighboring isotopes, the nature of the working material, the conditions under which the ion beam passes through the separating chamber, the compensation of the beam space charge, the required degree of enrichment of the isotopes, the extent to which the technological process has been developed for the particular element, etc.

The creation of high-capacity ion sources is perhaps the most difficult problem. In the process of developing and improving the method, the currents obtained from the ion sources were increased by factors of hundreds of thousands. At the present time, the currents of singly-charged ions reaching the receiver are approximately:

for light elements up to Mg	200-500 ma and above
for elements from Si to Zn	100-150 ma
for elements from Ga to Ba	50-100 ma
for rare earth elements and elements of the platinum group	10-50 ma

From this data we can evaluate the maximum possible accumulation of an isotope with a relative abundance C_0 over t hours continuous operation:

$$p = 0.0373 \, ItMC_0, \text{ g },$$

where I is the ion current at the receiver, amperes; t is the separation time, hours; M is the mass of the isotope, atomic units; and C_0 is the abundance, relative units.

Best Results of Isotope Separation on a Technical Scale

Element	Mass number of isotope	Content in natural element,%	Content in enriched element,%	Final form	Place of separation
Lithium	6 7	7,30 92,70	99,9 99,9	Li, Li_2SO_4, LiCl	1 1,2
Boron	10 11	18,83 81,17	>99 >99	B, H_3BO_3	3 3
Carbon	12 13	98,892 1,108	99,99 7,52	C	2 2
Magnesium	24 25 26	78,60 10,11 11,29	99,9 95,8 98,12	MgO	3 3 2
Silicon	28 29 30	92,18 4,70 3,12	99,4 68,6 64,0	SiO_2	2 2 2
Sulfur	32 33 34 36	95,018 0,750 4,215 0,017	98,45 9,8 33,6 0,88	S	2 2 2 1 2
Chlorine	35 37	75,4 24,6	92,4 65,6	AgCl	2 2
Potassium	39 40 41	93,08 0,0119 6,91	99,96 3,2 (7,75) 99,21	KCl	1,2 1 (2) 2
Calcium	40 42 43 44 46 48	96,92 0,64 0,132 2,13 0,0032 0,179	99,97 88 74 98 10,16 86,9	$CaCO_3$	1,2 3 3 3 2 1
Titanium	46 47 48 49 50	7,95 7,75 73,45 5,51 5,34	84,26 82,05 99,23 77,62 84,69	TiO_2	2 2 2 2 2

(Continued)

Element	Mass number of isotope	Content in natural element,%	Content in enriched element,%	Final form	Place of separation
Vanadium	50 51	0,24 99,76	28,6 99,98	V_2O_5	1 1, 2, 3
Chromium	50 52 53 54	4,31 83,76 9,55 2,38	92,3 99,6 96,8 88,95	Cr_2O_3	1,3 1 1 2
Iron	54 56 57 58	5,81 91,64 2,21 0,34	96,5 >99,9 87,29 86,0	Fe, Fe_2O_3	1 3 2 2
Nickel	58 60 61 62 64	67,76 26,16 1,25 3,66 1,16	99,3 99,2 80,9 97,7 98,5	NiO	1,2 3 2 3 1
Copper	63 65	68,94 31,06	99,7 99,4	Cu, CuO, $CuSO_4$	2 3
Zinc	64 66 67 68 70	48,89 27,81 4,07 18,61 0,62	98,3 96,9 71,2 95,7 48,4	Zn, ZnO	1 1 1 1 2
Gallium	69 71	60,16 39,84	98,42 98,08	Ga_2O_3	2 2
Germanium	70 72 73 74 76	20,55 27,37 7,61 36,74 7,67	88,1 89,2 68,9 95,2 79,3	GeO_2	2 2 2 2 2
Selenium	74 76 77 78 80 82	0,86 8,95 7,65 23,51 49,62 9,39	33,7 88,51 91,73 96,55 98,39 89,87	Se	1 2 2 2 2 2
Bromine	79 81	50,51 49,49	90,54 96,81	AgBr	2 2

(Continued)

Element	Mass number of isotope	Content in natural element,%	Content in enriched element,%	Final form	Place of separation
Rubidium	85 87	72,15 27,85	99,5 98,3	RbCl, Rb$_2$CO$_4$	1 1
Strontium	84 86 87 88	0,55 9,75 6,96 82,74	63,68 89,02 73,01 99,8	SrCO$_3$, SrC$_2$O$_4$	2 2 2 1
Zirconium	90 91 92 94 96	51,46 11,23 17,11 17,40 2,80	98,66 86,89 95,38 97,92 89,48	ZrO$_2$	2 2 2 2 2
Molybdenum	92 94 95 96 97 98 100	15,84 9,04 15,72 16,53 9,46 23,78 9,63	96,5 91,2 91,27 92,3 89,9 97,2 93,0	MoO$_3$	1 1 2 1 1 1 2
Silver	107 109	51,35 48,65	98,5 99,54	Ag, AgCl	1 2
Cadmium	106 108 110 111 112 113 114 116	1,22 0,98 12,35 12,78 24,00 12,30 28,75 7,63	39,2 41,6 70,0 64,5 (82,0) 86,9 54,1 97,0 81,0 (82,6)	Cd, CdO, CdSO$_4$	1 3 2 2 (1) 3 2 1 3 (1)
Indium	113 115	4,23 95,77	77,5 99,94	In, In$_2$O$_3$	1 1,2
Tin	112 114 114 116 117 118 119 120 122 124	0,90 0,61 0,35 14,07 7,54 23,98 8,62 33,03 4,78 6,11	75,4 57,2 37,0 97,8 85,6 94,91 79,82 99,2 96,3 96,3	Sn, SnO$_2$	1 1 1 1 1 2 2 1 2 1

(Continued)

Element	Mass number of isotope	Content in natural element,%	Content in enriched element,%	Final form	Place of separation
Antimony	121 123	57,25 42,75	99,4 96,7	Sb, Sb_2O_4	2 2
Tellurium	120 122 123 124 125 126 128 130	0,092 2,32 0,88 4,51 6,99 18,53 32,57 34,11	33,6 95,2 80,3 89,2 95,4 98,0 98,8 99,5	Te	1 1 1 1 1 1 1 1
Barium	130 132 134 135 136 137 138	0,101 0,098 2,42 6,59 7,81 11,32 71,66	27,50 13,5 76,4 89,2 84,6 78,8 99,52	$BaCO_3$	2 1 1 1 1 1 1
Lanthanum	138 139	0,089 99,911	0,597 99,96	La_2O_3	2 2
Cerium	136 138 140 142	0,193 0,250 88,48 11,07	29,97 21,9 99,67 95,9	CeO_2	2 1 1 1
Neodymium	142 143 144 145 146 148 150	27,13 12,20 23,87 8,30 17,18 5,72 5,60	98,1 91,0 93,45 88,1 96,4 97,2 98,7	Nd_2O_3	1 1 2 1 1 1 1
Samarium	144 147 148 149 150 152 154	3,16 15,07 11,27 13,84 7,47 26,63 22,53	72,13 81,63 79,3 73,01 74,09 96,0 96,05	Sm_2O_3	2 2 1 2 2 1 2
Gadolinium	152 154 155 156 157 158 160	0,20 2,15 14,68 20,36 15,64 24,96 22,01	(96,0) (88,4) (97,3) (95,5) (91,4) (97,4) (99,9)	Gd_2O_3	(1) (1) (1) (1) (1) (1) (1)

(Concluded)

Element	Mass number of isotope	Content in natural element,%	Content in enriched element,%	Final form	Place of separation
Hafnium	174 176 177 178 179 180	0,199 5,23 18,55 27,23 13,73 35,07	7,85 48,46 61,71 80,91 46,57 97,3	HfO$_2$	2 2 2 2 2 1
Tungsten	180 182 183 184 186	0,126 26,31 14,28 30,68 28,6	10,9 94,25 93,0 99,9 98,9	W, WO$_3$	1 2 1 1 1
Rhenium	185 187	37,07 62,93	85,38 98,22	Re	2 2
Mercury	196 198 199 200 201 202 204	0,146 10,02 16,84 23,13 13,22 29,80 6,85	8,44 79,11 73,09 91,3 71,9 98,3 89,17	Hg, Hg$_2$Cl$_2$	2 2 2 2 2 2 2
Thallium	203 205	29,46 70,54	(96,0) 98,7	Tl, Tl$_2$O$_3$, Tl$_2$CrO$_4$	1 1
Lead	204 206 207 208	1,37 25,15 21,11 52,38	27,0 (49,5) 81,0 (91,3) 66,8 (87,4) 96,4 (99,7)	Pb, PbSO$_4$	2 (1) 2 (1) 2 (1) 2 (1)
Uranium	235 238	0,714 99,28	28,1 (99,999)		1 (1) 2

The actual amount isolated from the compartments of the isotope receiver will be less than that calculated (sometimes it is only 35-50%), as the formula does not allow for the trapping factor of isotopes in the receiver and losses during their extraction and chemical purification.

Available Amounts of Enriched Isotopes

Electromagnetic apparatuses make it possible to obtain enriched isotopes in milligram and gram amounts with a single ion beam in one day. These amounts vary over a wide range and, in the first place, depend on the original concentration of the given isotope.

Some approximate data is given below on the normally available weight of isotope supplies in relation to C_0.

For work of great interest, the weight supplied may be increased considerably.

The supplies of highly enriched isotopes of the permanent gases are about 0.1 cm^3 at high enrichment and 1-10 cm^3 at 10- to 15-fold enrichment.

Original concentration of isotope C_0,%	Weight of normal supply, mg	Original concentration of isotope C_0,%	Weight of normal supply, mg
0.01-0.1	1- 10	10- 25	250- 500
0.1-1	10- 50	25- 80	500-1000
1-5	50-100	80-100	1000-1500
5-10	100-250		

Degree of Enrichment of Isotopes

The degree of enrichment of isotopes is usually characterized by the enrichment factor α, which equals the ratio of the relative content of the given isotope in the enriched product (C) and in the original (C_0):

$$\alpha = \frac{C}{1-C} : \frac{C_0}{1-C_0} .$$

In separating isotopes by the electromagnetic method, the enrichment factor attained in one stage is exceptionally high and reaches hundreds and even thousands.

The table gives a summary of the best results for 180 isotopes obtained on a technical separation scale on apparatuses in the USSR and on the so-called "calutrons" in the USA [1,2] and in England [3,4].

Data obtained on apparatuses in the USSR are designated by the numeral 1 in the sixth column of the table; results obtained in the USA (Oak Ridge, plant U-12) are given the numeral 2; and those in England (Harwell), the numeral 3. The numerals in brackets refer to separations on apparatuses with increased dispersions, or to cases where particular measures were taken to obtain a high enrichment.

For practical purposes one can arbitrarily divide the enriched isotopes into four groups according to the degree of enrichment:

	Enrichment factor α
Low enrichment	less than 20
Average enrichment	20-40
High enrichment	41-100
Very high enrichment	over 100

The isotopes are supplied in the form of the compounds shown in the table. Usually the content of traces of other elements in the enriched product does not exceed 0.5%. Higher requirements as regards the content of any impurity can be provided for beforehand.

LITERATURE CITED

[1] C. P. Keim, J. Appl. Phys. 24, 10 (1953).

[2] Stable and Radioactive Isotopes. Catalog and Price List of Oak Ridge National Laboratory (1952).

[3] Radioactive Materials and Stable Isotopes. Catalog No. 3, Isotope Division (Harwell, July, 1954).

[4] Stable Isotopes, Enriched by the Electromagnetic Method (Harwell, 1955).

[5] A. N. Nesmeianov, A. V. Lapitskii and N. P. Rudenko, The Preparation of Radioactive Isotopes (Moscow, 1954).

[6] Isotopes (prospectus), All-Union Reagents, Min. Chem. Ind. (Moscow, 1957).

ULTRAHIGH-TEMPERATURE ION SOURCE FOR ELECTROMAGNETIC SEPARATION OF ISOTOPES OF ELEMENTS IN THE PLATINUM GROUP

V. M. Gusev

An ultrahigh-temperature ion source has been constructed for electromagnetic separation of isotopes of elements in the platinum group: Pd, Pt, Ru and Ir. It can also be used for separation of isotopes of other refractory elements. All known compounds of elements of the platinum group are thermally unstable and cannot be used as operating materials in existing medium-temperature and high-temperature ion sources, in which the operating temperatures reach 800 and 1500°, respectively. In the present case, the choice of a working material was limited to pure elements, which have high melting points and low vapor pressures. To maintain a discharge in vapors of these metals in an ion source it is necessary to develop temperatures up to 2800°. Below we present the basic features of the construction of an ultrahigh-temperature ion source and some results obtained in the separation of palladium isotopes in a small electromagnetic separator.

Description of the Source

The basic feature in the design of this ion source is the use of electron bombardment for heating the discharge chamber inside of which there is a crucible containing the metal. The use of electron bombardment in an electromagnetic separator is especially convenient: the source operates in a vacuum, at a high positive potential, and in a strong magnetic field.

A schematic diagram of the ion source is shown in Fig. 1. The discharge 6, with a thermal cathode (indirectly heated) 2, is maintained in vapors of the element being separated in a gas discharge chamber 5 which looks like a truncated cone when viewed externally (diameter 20/15 mm , height approximately 75 mm). Inside the gas-discharge chamber there is a crucible 7 containing the metal; the capacity of the crucible is sufficient for approximately 10 hours of continuous operation with palladium, ruthenium or platinum. The metal in the crucible serves as the anode. The gas-discharge chamber is heated by bombarding the lateral surface of the graphite cone with a hollow cylindrical electron beam 8 at a voltage of 20-25 kv, which is collimated by the uniform magnetic field. The electron source is a gun consisting of a graphite sheet emitter 11 heated by direct current and a system of two diaphragms, with the first 10 at the potential of the emitter (ground) and the second 9 at a positive potential of 20-25 kev. The shape of the electron beam is determined by annular channels cut into these diaphragms. The front part of the cone is beveled and a slit 4 (40 × 2 mm) serves for extraction of the ions. The electrons are accelerated and the ions are extracted from the source by the same high voltage.

The ion-optical system of the source (Fig. 2) consists of two straight electrodes: the base electrode which is at ground potential, and an intermediate electrode, at a potential of 5 kv below the potential of the base electrode. The intermediate electrode sets up a potential barrier to prevent the loss of electrons which neutralize the space charge of the beam and which cause a high positive potential at the end of the source. The intermediate electrode is heated to a high temperature by direct current in order to prevent condensation of metal vapors in the region near the slit.

In order to reduce energy losses due to radiation, the cone and the electron beam are surrounded by a series of thermal shields. Through the use of a large number of shields the heat loss due to radiation can be reduced greatly. Because of structural considerations, the heat-shielding system of the discharge chamber consists of four graphite cylinders separated by gaps of 5-6 mm. With this system the power requirements for electron bombardment of the cone are reduced by a factor of five.

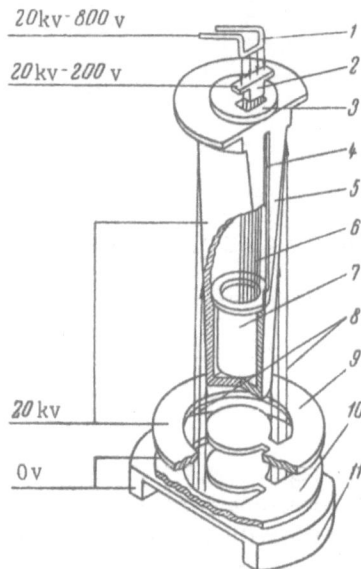

20kv-800 v — 1
20kv-200 v — 2
— 3
— 4
— 5
— 6
— 7
— 8
— 9
20 kv — 10
0 v — 11

Fig. 1. Diagram of the ion source: 1) filament for heating the cathode; 2) indirectly heated cathode; 3) cover for the cone; 4) slit for ion extraction; 5) gas discharge chamber (a truncated cone in the exterior view, a hollow cylinder inside); 6) discharge in the vapors of the element being separated; 7) crucible with metal; 8) 20-25 Kev bombarding electron beam; 9) and 10) diaphragms; 11) graphite sheet emitter.

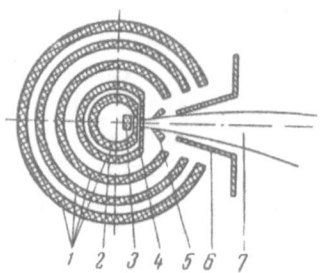

1 2 3 4 5 6 7

Fig. 2. Ion-optical system and heat shields in the source: 1) heat shields; 2) cone (gas discharge chamber); 3) discharge; 4) slit for ion extraction; 5) intermediate electrode; 6) base electrode; 7) ion beam.

One of the problems in this source is the suppression of parasitic high-voltage discharges which occur near the source. There are several types of discharges of this kind, but the most dangerous are cold Townsend discharges near the end of the source and trochoidal discharges, which can occur at any point at which the electric field E is perpendicular to the magnetic field H. In order to prevent the formation of a Townsend discharge, the end of the source is enclosed in two box-like covers with a small gap between them. The inner cover (made from graphite) is fastened directly to the end of the discharge chamber and is at a high potential while the outer cover (made from stainless steel) is attached directly to the supports of the base electrode and is at ground potential. With a sufficiently small gap between the covers (6 mm) and a low pressure in the vacuum chamber, the conditions in this gap correspond to the region to the left of the Paschen curve and the formation of a discharge becomes difficult. The second type of discharge is due to the trochoidal motion of electrons along the equipotential surfaces at the end of the source; this motion is accompanied by electron multiplication and an increase in the current to the source, which is at a positive potential. This situation leads to the dissipation of a sizeable part of the electrical energy and excessive local heating and, indeed, in certain cases, to the destruction of the metallic surfaces and the insulators. In order to prevent this type of discharge the box covers are provided with ribs ("labyrinth") by virtue of which a component of electric field is set up parallel to the magnetic field. The vertical component of the electric field tends to move single electrons to the positively charged cap of the source without multiplication. The rib system is used only on the left side of the source because all electrons which leave the beam move in this direction. Electron traps of this kind were first described in [1].

The primary construction material of the source is graphite. A photograph of the source is shown in Fig. 3.

Separation of Palladium Isotopes

The source described here has been used to separate palladium isotopes. The separation was carried out in a small 180°-electromagnetic separator. The radius of the mean ion projectory between the source and the detector is 55.5 cm. The distance between the focal points of two neighboring isotopes in the palladium beam (dispersion) is approximately 5 mm. Because of the small resolving power of the system, it is difficult to achieve high enrichment of the palladium isotopes. Using the highest possible magnetic field, the accelerating voltage for work with Pd is limited to approximately 20 kv. Preliminary work has been carried out to ascertain

Fig. 3. Ultrahigh-temperature ion source (the accelerating electrodes are removed).

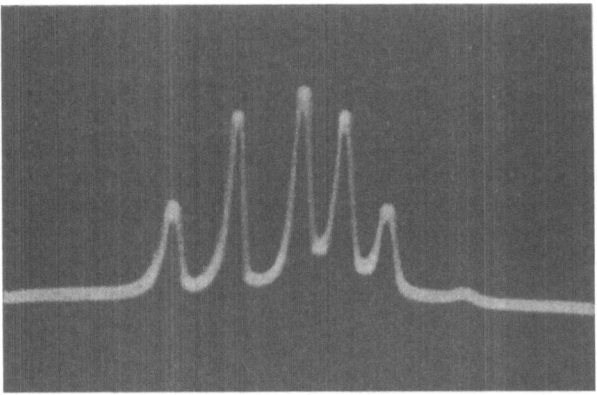

Fig. 4. Oscillogram of the isotope spectrum of palladium.

the mode of operation of the ion source most suitable for separation of palladium isotopes. The choice of the "working" mode of the source involves the optimum values of the vapor pressure in the discharge chamber, the discharge voltage and current, the accelerating voltage, and the vacuum in the separator chamber. The basic criteria for operation are the magnitude of the ion current at the detector and the quality of the beam focusing, which is controlled by the modulation factor of the ion current of the beam:

$$M = \frac{\Delta I^+}{I^+} \quad ,$$

where I^+ is the ion current in the beam and ΔI^+ is the alternating component of the beam current (measured with an oscilloscope or a vacuum-tube voltmeter).

Fluctuations in the beam depend chiefly on the conditions in the discharge. Studies which have been carried out indicate that in a source using a discharge which is heated in vapors of metallic palladium (the same as in vapors of other pure metals) there is a narrow region of values for the current, voltage, and pressure in the discharge for which the fluctuations in the beam are a minimum (M ≤ 1-2%), and the isotopic lines are well defined. Below we present data characteristic of the operation of the ion source when used for separation purposes.

Discharge current 0.5-0.6 amp.

Discharge voltage 200-250 volts.

Power required for electron bombardment (by which the vapor pressure in the discharge chamber was estimated), 1.5 kw.

Bombardment current 75 ma.

Accelerating voltage 20 kv.

Working vacuum in the separation chamber $3-5 \cdot 10^{-5}$ mm Hg.

In operation under these conditions the total current of palladium ions extracted from the source is 25-30 ma, while the current focused on the detector is 13-18 ma.

The six isotopes of palladium are separated simultaneously although it was difficult to instal six electrically insulated housings for the detector in a limited volume. Because of the low vapor pressure of palladium, its high melting point, and the small amount of power dissipated in the detector, it is not necessary to water cool the copper housings. The positions of the lines in the detector are monitored by the ratio of currents in the detector. During separation the beam focusing and resolution of the isotope lines is monitored with an oscilloscope. An oscillogram of the isotope spectrum for palladium is shown in Fig. 4.

At the end of the separation cycle the copper linings of the detector are treated with aqua regia. The palladium is precipitated from a weak base medium by a 1% alcohol solution of dimethylglyoxal. Approximately one gram of enriched palladium isotopes is obtained as a result of chemical processing. The average consumption of metallic palladium is about 0.8 grams/hour and the utilization factor, as determined from a measurement of the detector current, is about 7%.

An MS-3 mass spectrometer is used to analyze the enriched isotopes in the metallic samples [2]. Typical results are shown below.

Isotope	Content in a natural mixture, %	Content in enriched isotope sample, %	Isotope	Content in a natural mixture, %	Content in enriched isotope sample, %
Pd^{110}	13.5	58.1	Pd^{105}	22.6	53.7
Pd^{108}	26.8	84.3	Pd^{104}	9.3	46.3
Pd^{106}	27.2	77.4	Pd^{102}	0.8	31.5

This ultrahigh-temperature ion source provides temperatures at which a discharge can be maintained in vapors of platinum, ruthenium and iridium (about 2800°). The source provides a stable beam of platinum ions: a current of 5-7 ma is focused on the detector. In operation with platinum the power consumption in electron bombardment of the chamber is approximately 3.5 kw. The source also provides a beam of ruthenium ions (5 to 10 ma) with a power consumption of 5 kw for the electron bombardment.

The source described here has been used successfully for separation of small amounts of isotopes of other refractory elements; in particular, isotopes of silver, copper, boron, germanium, tin, titanium and nickel.

LITERATURE CITED

[1] S. E. Rauch, J. Appl. Phys. 22, 1128 (1951).

[2] K. G. Ordzhonikidze and G. D. Zubarev, Relative Abundances of Germanium and Palladium Isotopes, Present Collection, p.69.

INHOMOGENEOUS-FIELD MASS-SPECTROMETER
FOR ANALYSIS OF LIGHT-ELEMENT ISOTOPES

N. E. Alekseevskii, A. V. Dubrovin, G. I. Kosourov, G. P. Prudlovskii,

S. I. Filimonov, V. I. Chekin, V. N. Shelianin, and T. K. Shuvalova

As has been reported earlier [1], a mass-spectrometer with an inhomogeneous magnetic field has been designed and built at the Institute for Physical Problems, Academy of Sciences, USSR.

In contrast with ordinary mass-spectrometers, in which a homogeneous magnetic field is used, in these instruments use is made of an axially symmetric magnetic field which varies with radius in accordance with the relation

$$H = H_0 \left(1 - n \frac{r - r_0}{r_0} \right),$$

where r_0 and H_0 are the equilibrium radius and the magnetic field at the equilibrium ion trajectory

$$H_0 r_0 = c \sqrt{\frac{2 M u}{e}} \; ;$$

and $\underline{n}$ is the inhomogeneous magnetic field index

$$n = \frac{-\partial H}{\partial r} \cdot \frac{r}{H} \Big|_{r=r_0}, \ 0 \leqslant n < 1.$$

An ion beam which is paraxial about the equilibrium radius r_0 will be focused in the radial and axial directions after passing through angles equal correspondingly to

$$\Phi_r = \frac{\pi}{\sqrt{1-n}} \; ; \quad \Phi_z = \frac{\pi}{\sqrt{n}} \; .$$

The ion dispersion by mass (or energy) in a device of this kind is a factor of $1/(1-n)$ greater than the dispersion for an instrument of the same radius in which a uniform magnetic field is used. For the same slit widths in the source and detector, the resolving power is also increased by a factor of $1/(1-n)$.

A particular case which is frequently encountered in practice is an instrument with a sectored magnetic field, i.e., instruments in which the ions move in a magnetic field for only part of the trajectory. For instance, if we use an ion deflection angle of $2\psi = 180°$ in the magnetic field and locate the source and the detector at equal distances from the boundaries of the magnetic field f, focusing is achieved when

$$f = \frac{r_0}{\sqrt{1-n}} \operatorname{ctg} \psi \sqrt{1-n}.$$

In this case the formulas for dispersion and resolving power remain the same as for a nonsectored magnetic field.

The use of an inhomogeneous magnetic field in a mass-spectrometer results in a number of features which open new possibilities for this instrument as compared with the uniform-field version.

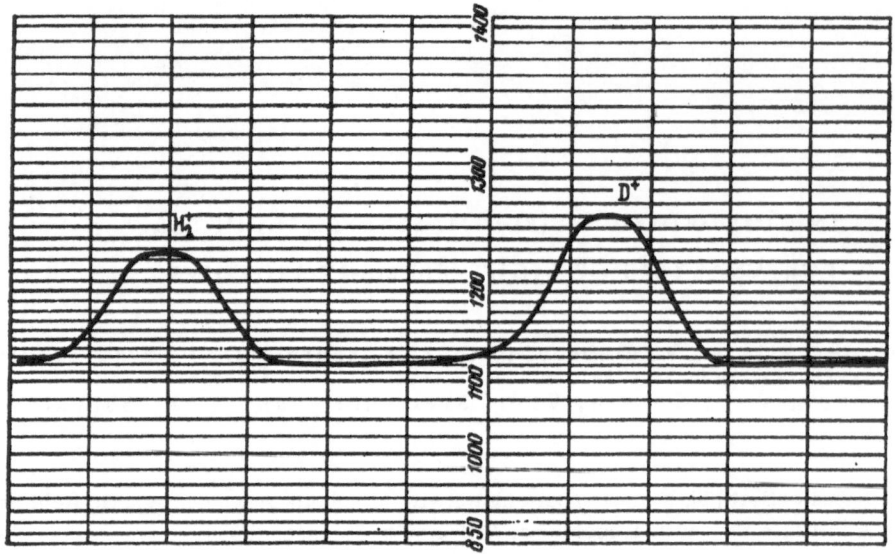

Fig. 1. The mass doublet $2H_2^+ - D^+$.

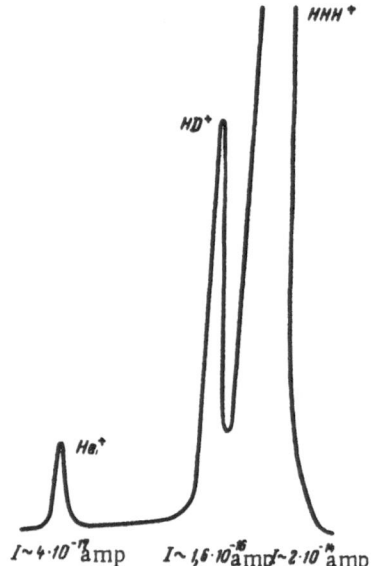

Fig. 2. The mass triplet $3He^{3+} - HD^+ - HHH^+$.

1. It has been found possible to increase the dispersion and resolving power by a factor of 5-9 without increasing the radius of the instrument or the size of the magnet. No reduction in slit widths is required; consequently, the increase in resolving power is achieved without any deterioration in the sensitivity of the instrument.

2. For a given sensitivity and resolving power a considerable reduction in the size of the instrument can be achieved.

3. Using a comparatively small radius and a reasonable resolving power, it is possible to increase the sensitivity of the device by using wider slits in the source and detector.

It should be noted that the use of an inhomogeneous field means that there can be a considerable reduction in the ion path between the source and the detector (approximately a factor of two as compared with a device with a deflection angle of 180° and almost a factor of three in a device with a deflection angle of 60° with the same dispersion).

The instruments which have been designed and developed at the Institute for Physical Problems in which the above principles are used have been built chiefly to obtain high resolving power and increased sensitivity. All of these instruments are characterized by ion—beam deflection angles of 180°. Some typical data are given below.

A glass instrument is characterized by a radius r_0 = 40 mm, n = 0.89, f = 203 mm, source slit width 0.2 mm, and detector slit width 0.4 mm [1].

The experimentally obtained resolving power for line width at half height is approximately 700, which is sufficient for complete resolution of the mass doublet $4He^+ - D_2^+$.

Another instrument is an all-metal mass spectrometer with radius r_0 = 350 mm, n = 0.87, f = 1400 mm, which has a dispersion of 27 mm for a 1% change in mass [1]. Using detector and source slit widths of 0.2 mm a resolving power of approximately 7500 has been obtained (half width).

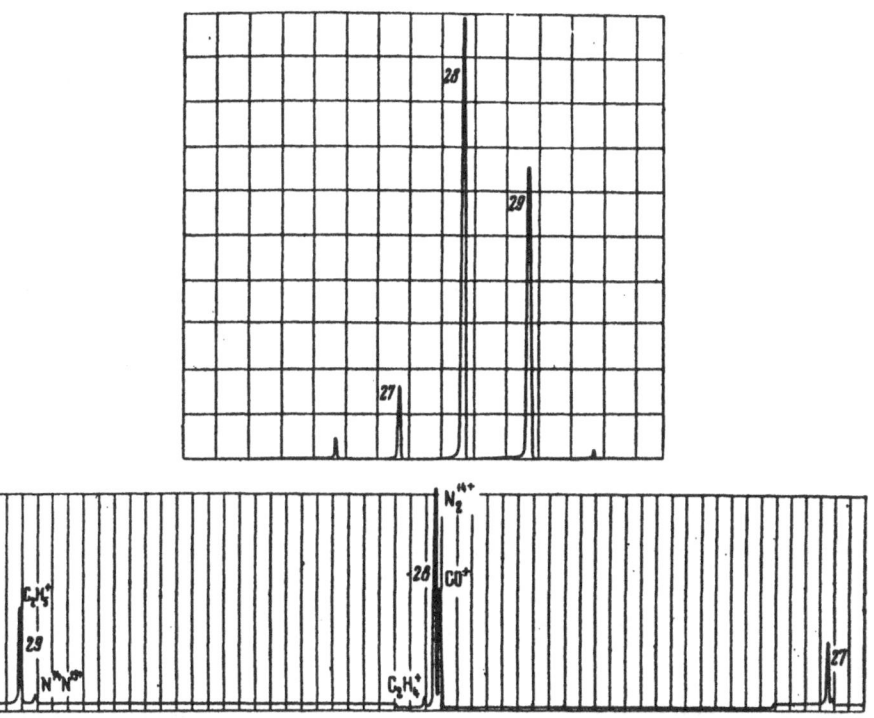

Fig. 3. The spectrum at mass numbers 27 – 28 – 29: a) mass spectrometer with uniform field (MS-4); b) mass spectrometer with inhomogeneous field (IFP).

It is apparent from the mass spectrograms that have been presented in [1] that all lines are completely resolved. Regardless of the length of the ion path, the instrument has a rather high transmission; operating at a source gas pressure of approximately 10^{-5} mm Hg, the ion current at the collector is greater than 10^{-9} amp.

In another instrument with radius r_0 = 152 mm, it has been possible to achieve a resolving power of 11,000 [1,2] with detector and source slit widths of approximately 0.05 mm.

The instruments which have been built, which are characterized by deflection angles of 180°, are not the only possible means of applying an inhomogeneous magnetic field in the mass spectrometer. It is also possible, for example, to build an instrument in which the ion beam turns through an angle of 270° in the magnetic field; in this case a considerable increase in dispersion can be achieved with a relatively small arm length f. Finally, it is possible to build devices in which the ion beam is focused after completing several revolutions in the magnetic field.

The instruments described above, in which an inhomogeneous magnetic field is used, are especially suitable for isotope analysis of light elements. In this case, because of the low mass, the radius of the instrument can be rather small whereas a high resolving power, required for resolving multiplets, can be achieved because of the increased dispersion.

Instruments built by us have been used successfully in isotope analysis of light gases. In analyzing hydrogen for deuterium content, as is well known, it is important to resolve the mass doublet $3HHH^+ - HD^+$ (when the deuterium content is small). In this case, because of the high resolving power, it is possible to carry out an analysis in one measurement, without requiring a difficult and highly inaccurate pressure analysis.

In an analysis of mixtures containing 90-95% deuterium the use of high resolving power affords the possibility of resolving the doublet at mass 2 (Fig. 1). The advantage here can be illustrated by the following example. In carrying out an analysis of a D – H mixture an isotopic ratio of $H_2/D_2 = 3.46 \pm 0.05\%$ was obtained (i.e., the relative error was less than 1.5%). This analysis was repeated over the course of a three-day period and the reproducibility of the results was good. If the same analysis was carried out with an instrument of low resolving power, the doublet at mass 2 could not be resolved. If the peak at mass 2 is assigned to the H_2^+ ion, then the relative error $C_H = H_2/D_2$ is increased to 12%; if this line is assigned to the D^+ ion, the error in C_H is increased to 20%.

High resolving power is required in order to analyze helium for small He^3 content. In this case it is very important to resolve the mass triplet $3He^{3+} - HD^+ - HHH^+$ inasmuch as the hydrogen peaks in the residual gases may cause a considerable distortion in the results of the measurement. The use of a high resolution device makes it possible to avoid this danger completely. In this case the presence of the background hydrogen lines in the mass spectrum does not hinder the analysis but actually makes it easier since it affords a means of easy location of the weak He^3 line in the mass spectrum (Fig. 2). It should be noted also that in this case the time required for preliminary processing of the instrument can be reduced considerably.

At He^3 concentrations of approximately 10^{-4} the relative error is less than 1%. At He^3 concentrations less than 10^{-6} the error is less than 10^{-7}, indicating a relative error smaller than $\pm$ 10%.

It may also be pointed out that the sensitivity obtained in these devices is such that the ion currents can be measured with ordinary vacuum-tube voltmeters.

The high resolving power of these devices has also been found extremely useful in making isotope analyses of nitrogen. As is apparent from Fig. 3b, lines 28 and 29 have a complicated structure and to obtain accurate data from an analysis of nitrogen it is necessary either to purify carefully the nitrogen sample being analyzed from hydrocarbons and oxygen and to bake out the instrument carefully, or to use a high resolving power instrument.

The high-resolution instruments also make it possible to analyze the isotopic composition of neon in the presence of argon. These instruments have been successfully used for a number of gas–chemical analyses [3].

An inhomogeneous field has also been used to increase the sensitivity of the mass spectrometer. An instrument has been built and successfully used in the laboratory in which both high sensitivity and high resolving power are obtained.

One of these instruments (glass) is characterized by the following parameters: r_0 = 50 mm, source slit width 0.8 mm, and detector slit width 2 mm. The resolving power for half width is approximately 120.

A collector current of approximately 10^{-8} amp is easily obtained with this instrument, making it possible to measure concentrations of 10^{-6} and lower.

The high sensitivity means that it is possible to operate at low source pressures, thereby reducing the number of HHH^+ ions and increasing the accuracy of hydrogen analyses at low deuterium concentrations.

A similar glass instrument with a radius of 40 mm and a resolving power of approximately 200 has been used for analysis of very small amounts of noble gases. This instrument was filled with the gas being analyzed at a pressure of 10^{-5}-10^{-6} mm Hg and sealed off. The results which were obtained indicate that it is possible to analyze 10^{-9} cm^3 of helium with this instrument. A further improvement in resolving power to 500 and the use of a secondary electron multiplier would make it possible to increase still further the sensitivity of such a device in order to analyze small amounts of gas.

From what has been indicated above, it is apparent that the application of an inhomogeneous magnetic field in the mass spectrometer has a number of advantages as compared with uniform-field devices and will make it possible to solve a number of problems relating to isotope analysis of gases.

LITERATURE CITED

[1] N. E. Alekseevskii, G. P. Prudkovskii, G. I. Kosousov and S. I. Filimonov, Proc. Acad. Sci. SSSR, 100, 229 (1955).

[2] A. V. Dobrovin and G. V. Valabina, Proc. Acad. Sci. SSSR, 102, 719 (1955).

[3] N. E. Alekseevskii, V. L. Tal'roze and V. N. Sheliapin, Proc. Acad. Sci. SSSR 93, 997 (1953).

THE RELATIVE ABUNDANCE OF PALLADIUM AND GERMANIUM ISOTOPES

K. G. Ordzhonikidze and G. N. Zubarev

Palladium

There are great discrepancies in the published data on the isotopic composition of palladium [1,2]. The accuracy of Sampson and Bleakney's measurements [1] is low in comparison with what may be obtained on modern mass spectrometers. In addition, the mass spectrogram they presented shows incomplete resolution of the palladium isotopes, which undoubtedly lowers the value of the results. Only the results of measuring the isotopic composition of palladium were given in a short report in 1953 [2]. Nothing is said on the accuracy of the measurements and the method in the report.

The aim of the present work was to develop a method of determining the isotopic composition of natural palladium accurately and of measuring the degree of enrichment of a palladium sample, obtained on an electromagnetic separating apparatus. For this, the amount of palladium required for the isotopic analysis could not exceed 1-2 mg.

Due to the low vapor pressure of metallic palladium, we had to use an ion source with a thermal shield or develop some type of evaporator, which made it possible to work at a low enough heater power to avoid overheating of the ion source elements. We were able to prepare such an evaporator by using the MS ion source, shown schematically in Fig. 1, without any modification.

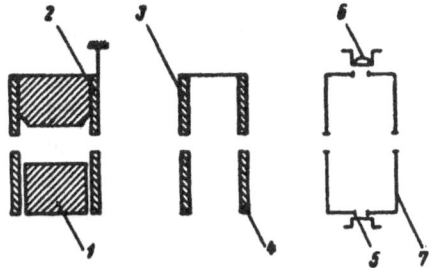

Fig. 1. Plan of ion source: 1) deflecting electrode; 2) accelerating lens; 3) extracting lens; 4) focusing lens; 5) cathode; 6) evaporator; 7) anode chamber.

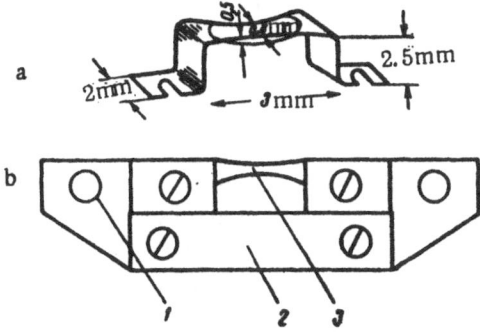

Fig. 2. Evaporator: a) heater; b) holder; 1) openings for fixing; 2) ceramics; 3) heater.

The evaporator consisted of a tungsten or tantalum strip, 30 μ thick. The form and dimensions of the evaporator are shown in Fig. 2a. The special form of the strip made it possible to produce more intense heating at that part of the evaporator where a hollow had been made with a die. The substance investigated was placed in the hollow. The evaporator was fixed in a holder (Fig. 2b), which was mounted in the ion source with screws through the two holes 1 (Fig. 2b).

The evaporator was heated by passing a current through it. A stable and intense palladium ion beam was obtained with an evaporator heater power of 15-20 watts. This produced no noticeable heating of the ion source elements, including the ionization chamber. When the evaporator was loaded with a powdered mass of palladium, a more stable ion beam was observed than when pieces of metal were used. This may be explained by the fact that in the first case there was better thermal contact between the investigated material and the evaporator and more uniform evaporation. The capacity of the evaporator was about 2 mg. This evaporator was convenient in operation and made it possible to measure ion currents of palladium isotopes of the order of 10^{-10}-10^{-11} amp at ion source pressures not exceeding $3 \cdot 10^{-7}$-$5 \cdot 10^{-7}$ mm Hg, which prevented the appearance of secondary effects, that arise at elevated pressures.

The most stable ion beams were obtained with tungsten powder added to the material investigated. Mixing tungsten powder with the material investigated stabilized the evaporation process, as in this case the material investigated did not evaporate immediately; its vapor diffused only gradually through the pores of the spongy, sintered tungsten powder. Therefore, the evaporation process lasted longer and due to this it was possible to analyze even the small amounts which could not be investigated without the addition of tungsten powder. We should also note that we used this type of evaporator with great success for the analysis of zinc, germanium and solid compounds of boron. In the evaporation of solid compounds of boron ($Al_2O_3 \cdot KBF_4$), the addition of tungsten powder played a decisive role in the production of stable ion beams of boron and its compounds.

The isotopic analysis of palladium was performed on an MS-3 mass spectrometer, which had the same parameters as the apparatus developed in our laboratory [3].

Ion beams of palladium isotopes began to appear at a temperature of 1260-1300°, and then the intensity of the palladium ion beam rapidly increased until the DC amplifier cut off. To obtain measurable ion beams, the temperature of the evaporator had to be reduced to 1100°. At this temperature the intensity of the ion beam remained constant for some time, depending on the amount of palladium investigated. For example, with an evaporator load of 1 mg of palladium, the intensity of the ion beam remained constant for 5 hours. Then the ion beams of the palladium isotopes began to fall. Increasing the temperature of the evaporator to 1800-1900° was then inadequate for maintaining the intensity of the ion beam. At the moment when the temperature was raised, the intensity of the beam increased slightly, but it immediately fell rapidly.

The rapid increase in the intensity of the palladium ion beam at a temperature of 1300° was accompanied by a considerable (by 1.5 to 2 orders) decrease in the intensity of the mercury mass lines in the apparatus. With the fall in the intensity of the palladium ions, the intensity of the mercury ions was observed to increase gradually until the original value was reached. These phenomena are characteristic of palladium. They were not observed in the analysis of other metals. The behavior of the residual gases and radicals – CO, CO_2, HO, H_2O and O_2 – was the same as in the analysis of other metals. The intensity of the mass lines of CO, CO_2, H_2O and HO remained almost the same as before the appearance of ions of the metal investigated and the O_2 line decreased by only 20%.

The phenomena characteristic of the evaporation of palladium may be explained by the interaction of mercury vapor with the incandescent metal surface. Apparently, at a temperature of about 1300° a palladium amalgam is formed and as a result of this the vapor pressure of the palladium may increase and that of the mercury decrease. Therefore, the intensity of the palladium ions increases and sufficient palladium evaporates at lower temperatures (1100°) for measurement of the ion beam intensity. The consumption of mercury in the formation of the amalgam and the decrease in its vapor pressure causes the decrease in the mercury ion intensity. When the palladium disappears from the evaporator, there is a gradual restoration of the mercury ion intensity to the original value.

The possibility of forming a palladium amalgam was reported as early as 1852 [4]; some properties of natural compounds of the amalgam type are given in Gmelin's handbook [5]. No data has appeared in the literature up to now on the phase diagram of the Pd – Hg system.

The interaction of mercury vapor and palladium and its effect on the yield of palladium ions, which we discovered, is of great interest and requires further investigation to determine the dependence of the ion yield on the amount of mercury vapor striking the palladium surface.

To study the systematic errors of mass spectrometric measurements, we performed experiments with the inert gases, mercury and heavier masses on a laboratory mass spectrometer. As a result of these experiments, it was established that with correct alignment of the apparatus (complete resolution of the isotopes registered, sharp, symmetrical lines and a low pressure in the ion chamber – not more than 10^{-6} mm Hg), the mass spectrometer was free from systematic errors within the limits of the statistical errors. Then the reproducibility of the

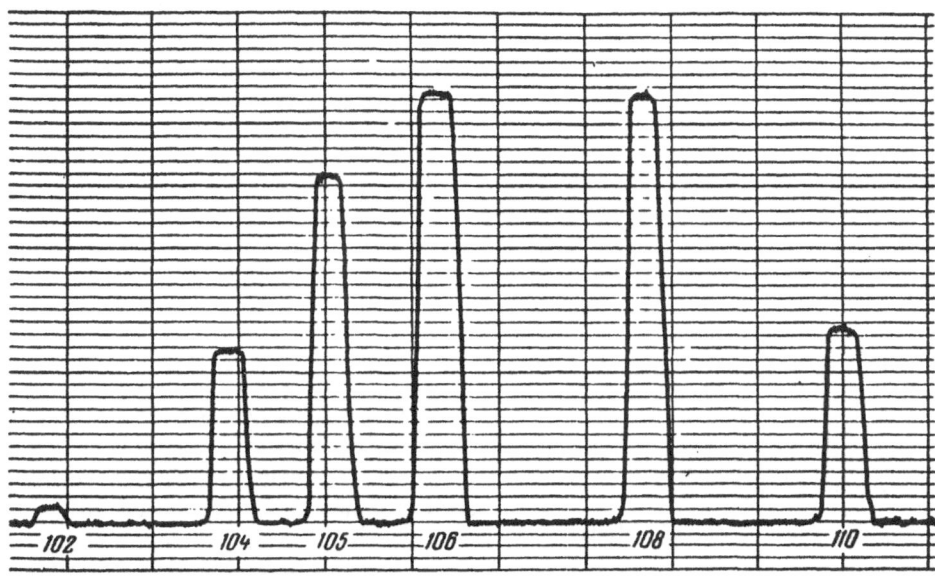

Fig. 3. Mass spectrogram of palladium isotopes.

measurement results did not depend on the time interval between measurements. We should note that in those cases where separating effects in the ion sources or in the input systems were observed, large errors could result in determining the isotopic composition of the elements investigated, especially with light ones. In these cases it was necessary to study the separate processes carefully and to introduce reliable corrections into the measured values. This type of work was performed for lithium [3] and potassium [6]. In the case of palladium no separating processes were observed in the ion source during the evaporation process, right up to the disappearance of the material from the evaporator. The measured values of the relative isotopic composition remained constant within the limits of the experimental errors during the whole experiment. The investigations were performed on an MS-3 mass spectrometer, which was adjusted according to singly- and doubly-charged mercury ions. The results of measuring the isotopic composition of mercury were in complete agreement with the measurements carried out by the authors of the present article on a laboratory mass spectrometer.

The mass spectrogram of palladium isotopes is presented in Fig. 3. The palladium isotopes were recorded on a Speedomax Recorder with linear scanning of the magnetic field. As is shown, the palladium isotopes were resolved completely and the mass lines recorded had flat tops.

The ion currents of the palladium isotopes were measured by an absolute method — the collection of all six isotopes on one collector. To obtain the same measurement accuracy in recording the low-abundance isotopes with mass numbers 102, 104 and 110, the sensitivity of the recording instrument was increased according to their abundance.

The relative abundance of palladium isotopes was determined on singly-, doubly- and triply-charged palladium ions. In the presence of even low intensities of mercury ions, the superposition of doubly-charged Hg^{204} ions on the singly-charged Pd^{102} ions was allowed for using the ratio $Hg^{202}/Hg^{204} = 4.43 \pm 0.01$, which we measured. The other mass lines of singly-charged palladium isotopes were free from the superposition of background lines.

At some mass lines of doubly- and triply-charged palladium ions with mass numbers 53, 54, 55, 35 and 36, we observed a background from residual gases, whose intensity was from 0.6 to 5% of the intensity of the measured mass line.

The results of measuring the isotopic composition of palladium on singly-charged ions are presented in Table 1. The maximum ionization potential of Pd^+ ions is 90 v. For comparison, the data from previously published papers are presented in the table.

TABLE 1

Palladium isotopes (mass numbers)	Isotope content,%		
	Data of authors	Sampson and Bleakney [1]	Communication of 1953 [2]
102	1,01±0,004	0,8	0,96
104	10,96±0,03	9,3	10,97
105	22,13±0,04	22,6	22,23
106	27,18±0,02	27,2	27,33
108	26,70±0,03	26,8	26,71
110	12,00±0,03	13,5	11,81

TABLE 2

Palladium isotopes (mass numbers)	Content in %	
	on Pd^{++} ions	on Pd^{+++} ions
102	0,97±0,007	0,96±0,01
104	10,85±0,04	11,03±0,05
105	22,11±0,04	22,36±0,08
106	27,24±0,02	27,03±0,03
108	26,73±0,04	26,65±0,08
110	12,09±0,04	11,98±0,08

As Table 1 shows, the accuracy of our values for the relative contents of palladium isotopes was 0.1-0.4%. The values we obtained for the palladium isotope composition do not agree with the previous results of Sampson and Bleakney [1], but are in quite good agreement with the data of 1953 [2]. The greatest deviation (about 1.5%) was obtained in determining the content of the palladium isotope with mass number 110.

The results of measurements on doubly- and triply-charged palladium ions are presented in Table 2. The maximum ionization potential of doubly-charged palladium ions is 110 v and of triply-charged palladium ions, 150 v. The intensity of the Pd^{++} ions obtained was a factor of three less and the intensity of the Pd^{+++} ions a factor of thirty less than the intensity of the singly-charged palladium ions. The accuracy of measurement was also correspondingly lower.

Comparison of the results presented in Tables 1 and 2 shows that the small superpositions, which occurred with the doubly- and triply-charged palladium ions, give small errors. Apart from the least abundant palladium isotope, the results on doubly- and triply-charged ions agree with the results obtained on singly-charged ions with an accuracy of 0.5-1%.

Using the method of adding tungsten powder to the substance investigated, which we developed, we were able to analyze small amounts of palladium isotopes — 0.1-0.03 mg. Here the ratio of the amount of tungsten to the amount of palladium was always greater than unity (from 3:1 to 5:1). The evaporation process, during which it was possible to perform an isotopic analysis on such small amounts as 30-40 μg, lasted for about 15 min. The accuracy in these cases was a factor of two to three worse than that given in Table 1.

The method we developed for the isotopic analysis of small amounts made it possible to perform an analysis on not only the palladium isolated chemically from the pockets of an electromagnetic separating apparatus, but also the material scraped directly from the surface of the pockets, without any chemical treatment. In these samples there was 10-50 times more copper (the material of the pockets) than palladium.

Germanium

The isotopic composition of germanium was determined by various authors [7-12], using halogen compounds of germanium. The results obtained in 1947-1953 agreed with each other with an accuracy of 1-2%.

The results of our measurements of the relative isotopic abundance of germanium isotopes, obtained on GeF_3^+ ions using germanium tetrafluoride and data from papers published previously, are presented in Table 3. The mass spectrogram which we plotted for GeF_3^+ ions, presented in Fig. 4, shows that the germanium isotopes were resolved completely.

TABLE 3

Germanium isotopes (mass numbers)	Content, %, according to data of					
	present work, GeF_3^+	Aston	Hibbs, Redmond, Gwinn and Harman		Inghram, Hayden and Hess Ge^+	Reynolds GeF_3^+
			GeF_s^+	GeF^+		
70	20,55±0,03	21,2	20,60±0,06	20,65±0,04	20,55	20,52±0,17
72	27,35±0,02	27,3	27,38±0,08	27,43±0,02	27,37	27,43±0,21
73	7,78±0,04	7,9	7,83±0,06	7,86±0,04	7,67	7,76±0,08
74	36,50±0,04	37,1	36,40±0,10	36,34±0,05	36,74	36,54±0,23
76	7,86±0,03	6,5	7,78±0,05	7,72±0,01	7,64	7,76±0,08

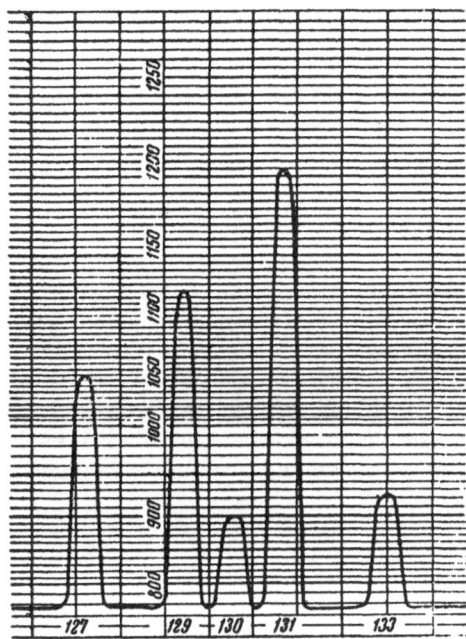

Fig. 4. Mass spectrogram of germanium isotopes, obtained on GeF_3^+ ions.

The deviations in the data in Table 3 are irregular. The results of our measurements agree with the results of Reynolds, within the limits of error of his measurements, which are half an order higher than ours. Neither the accuracy of the values nor the method of obtaining the ions are given in the data of Inghram, Hayden and Hess [9]. These data appear in tables of isotopes up to the present time.

The preparation of gaseous germanium halogen compounds involves a definite cycle of chemical reactions which, as is known, may be accompanied by isotope separation. Therefore, these reactions necessitated a careful control on the constancy of the isotopic composition or a 100% yield. The fulfillment of these requirements involved certain difficulties and often required a careful and lengthy processing of the samples measured, in particular, of small amounts of analysis sample. Therefore, mass spectrometrists have the problem of the isotopic analysis of metallic germanium or its oxide without any chemical processing. It also occurs in investigating the relative abundance of germanium isotopes in nature.

Germanium ions were obtained from germanium metal by Kohl [11] in determining impurities in germanium analysis samples. Judging by the tables presented in [9] and [11], Inghram, Hayden and Hess also performed an isotopic analysis of germanium on metallic germanium. As was indicated above, the method by which the latter authors obtained their ions is not known, but Kohl evaporated metallic germanium from a quartz crucible. According to him, the preparation from an evaporator of any metal is excluded due to germanium's capacity for forming alloys with metals readily. Apparently this also explains the fact that metallic evaporators have not been used for germanium up to the present time.

Using an MS ion source (see Fig. 1) we performed experiments on the isotopic analysis of metallic germanium by evaporating it from a quartz crucible. These experiments showed that despite previous burning out of the furnace and the crucible, the background lines at the Ge^+ and GeO^+ ions did not disappear. With an increase in the crucible temperature, the intensity and the amount of background lines increased. The background lines observed were partly due to gas liberated from the strongly heated parts of the ion source, especially from the ionization chamber. Deposits were formed on the anode chamber and the first electrostatic lens, which upset the operating conditions of the ion source; their removal required the partial dismantling of the ion source.

These difficulties were overcome by using the tungsten or tantalum evaporator developed for palladium. The metallic germanium was placed in the evaporator, as in the case of palladium. However, in the case of germanium, the addition of tungsten powder played a greater role than for palladium. With the evaporation of pure germanium, the ion currents were not sufficiently stable for accurate measurements. The scattering of the isotope content values measured reached 3-4%. The addition of tungsten powder considerably stabilized and prolonged the evaporation process. The best ratio of tungsten-to-germanium was 3:1 to 5:1. Measurable germanium isotope currents of about 10^{-10} amp were obtained at a temperature of 920°. At this temperature the intensity of the germanium ions remained constant for some time and then began to fall. A constant intensity of the germanium ion beam was maintained by gradually raising the evaporator temperature to 1400°. In this respect, germanium behaved as other metals. The characteristic phenomena, observed in the evaporation of palladium, were not observed in the case of germanium.

By the method of adding tungsten powder to the material investigated, which was developed, isotopic analyses were performed on such small amounts of germanium as 0.2-0.1 mg. Measurements of the ion currents of germanium isotopes could be carried out for 1.5 to 2 hours. We should note than an x-radiogram of a mixture of powdered tungsten and germanium placed in the evaporator and heated to various temperatures (from 920 to 1500°) showed the presence of two separate components: tungsten with a constant lattice of 3.1648 ± 0.0002 A, and germanium with a constant lattice of 5.6570 ± 0.0002 A. The formation of alloys of germanium and tungsten was not observed.

The results of measuring the relative abundance of germanium isotopes are given below.

Germanium isotopes (mass numbers)	Content, %
70	20.33 ± 0.1
72	27.28 ± 0.08
73	7.89 ± 0.06
74	36.49 ± 0.08
76	8.01 ± 0.08

As is shown, the accuracy of measurement of germanium isotope ion currents with the use of gaseous GeF_4 is higher than with the evaporation of metallic germanium. Therefore, for precision measurements, it is advantageous to use gaseous GeF_4, while the use of metallic germanium may be recommended for the isotopic analysis of small amounts and for the measurement of a series of samples.

We carried out the isotopic analysis of a germanium sample enriched on an electromagnetic separating apparatus, using both germanium tetrafluoride and metallic germanium.

LITERATURE CITED

[1] M. B. Sampson and W. Bleakney, Phys. Rev. 50, 732 (1936).

[2] Bull. Amer. Phys. Soc. 28, 5, 24 (1953).

[3] K. Ordzhonikidze and V. Shiuttse, J. Exptl.-Theor. Phys. 29, 4 (10), 479 (1955).

[4] Hosford, Amer. J. Sci. 2, 13, 305 (1852); P. Gasmajor, Chem. News 34 (1876).

[5] Gmelins Handbuch der Anorgan. Chemie, S. N. 68, Platin (A), 285, 286 (1949).

[6] K. Ordzhonikidze and G. Dolidze (in press).

[7] R. F. Hibbs, J. W. Redmond, H. R. Gwinn and W. D. Harman, Phys. Rev. 75, 533 (1949).

[8] R. P. Graham, J. Macnamara, I. H. Crocker and R. B. MacFarlane, Can. J. Chem. 29, 89 (1951).

[9] M. G. Inghram, D. C. Hess and R. J. Hayden, Quoted in Table of Isotopes: Hollander, Perlman, Seaborg, Rev. Mod. Phys. 25, No. 2, 504 (1953).

[10] J. H. Reynolds, Phys. Rev. 90, 1047 (1953).

[11] G. Köhl, Z. Naturforsch. B. 9a, No. 11, 913 (1954).

[12] Aston, Mass Spectra and Isotopes [Russian translation] (Moscow, 1948) p. 166.

SOME PROBLEMS IN THE THEORY OF ISOTOPE SEPARATION

A. M. Rozen

A series of papers are devoted to the theory of the separation of isotopes on columns [1-6]. In the present article, the operation of columns and cascades of columns as a whole* are examined by a comparatively simple graphoanalytical method and some results obtained by other authors [4,5] are developed. An examination is made of cascades of distillation or exchange columns (with chemical phase transition), operating at any concentrations of the components extracted and including columns combining concentration with extraction. An examination is also made of the case of "desorption" phase transition [4] — enrichment by the two-temperature method (in single columns and in a cascade).

The first part of the work is devoted to the stationary states of cascades and columns, and the second to nonstationary states. In the first part, the mass transfer and material balance equations are solved in the integrated form, which makes it simpler to obtain the required results (to solve any actual problem it is sufficient to find the coordinates of the intersection of the operating line and the equilibrium curve). In the second part, together with integration of the nonstationary mass transfer equations, methods are developed based on the principles of the quasistationary state and similarity [5]. In all cases, the aim was to obtain results with a precise physical meaning and of sufficient simplicity for engineering application, but with a definite degree of rigidity of the basic premises and calculations.

I. Stationary State of a Cascade of Distillation or Exchange Columns

1. Material Flow in the Cascade, x — y Diagram

The cascade scheme is presented in Fig. 1, which also indicates all the flows and concentrations (the evaporators and other devices for phase transition are not shown). As in [3-5], it is assumed that the cascade is intended for concentration of the less volatile component. The extraction occurs in the upper part of the first column and concentration in the lower part, and in all the subsequent columns. The values considered as given are as follows: the amount of the starting mixture L_0, its concentration x_0 (where $x_0 \ll 1$), the output of the cascade $\underline{j}$, and the concentration of the final product x_p (the output as regards the intermediate product $P = j/x_p$). The number of stages $\underline{n}$, and the degree of separation in each column $q = \dfrac{x_s}{x_h} \dfrac{1 - x_h}{1 - x_s}$ (where x_s and x_h are the concentrations in the still and the head of the column) are normally accepted on the basis of previous calculations. We will also assume these values to be given. The degree of fractionation and material flows in the column were to be determined. For this purpose, let us plot the $x - y$ diagram of the cascade. Let us assume that the solution is ideal, i.e., the equilibrium composition of the gas (y_e) and the liquid phase (x) are related by Raoult's law.

For the less volatile component we have (Fig. 2),

$$y_e = \frac{x}{\alpha - (\alpha - 1)x} ,$$
(1)

where $\alpha = $ const. At small concentrations of one of the components, expression (1) becomes Henry's law:

$$y_e = x / \alpha.$$
(1a)

---

*In other words, an examination is made of columns ("straight cascades") or cascades with elements with a multiple separating action ("cascades of cascades"). However, some of the results are applicable to cascades with single-acting elements.

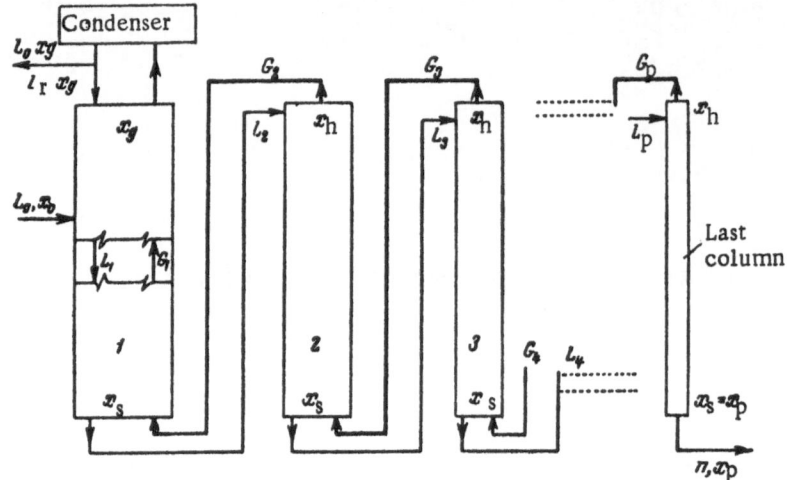

Fig. 1. Cascade scheme: L_0 — amount of mixture supplied for separation; L_r — amount of reflux; L_i and G_i — liquid and gas flows in the i-th column; P — the amount of final product; and x_0, x_d, x_h, x_s and x_p — the concentrations of the extracted component in the starting mixture, the distillate [$x_d = x_0 (1 - D)$], the head of the column, the still of the column and the final product.

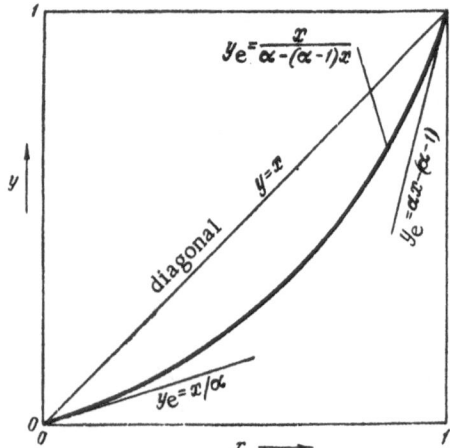

Fig. 2. Equilibrium for the low-volatility component.

It follows from the material balance that the equation of the operating line of the extraction (upper) section of the first column has the usual [7] form:

$$y = \frac{R_r}{1 + R_r} x + \frac{x_g}{1 + R_r} = \lambda x + (1 - \lambda) x_d, \quad (2)$$

where $R_r = L_r / L_0$ — the reflux ratio; $\lambda = L/G$ — the ratio of the liquid and gas flows; and x_d is the concentration of the combined distillate.

For all concentration columns, the material balance ($Lx - Gy = j$, $L - G = P = j / x_p$) gives

$$y = \frac{x - j / L}{1 - j / Lx_p}, \quad (3)$$

and when $x = x_p$, $y = y_p$, i.e., the corresponding point lies on the diagonal. For concentrating columns operating at $x \ll 1$, $j \ll L$, so that $L = G$

$$y = x - j/L, \quad (3a)$$

i.e., the operating lines are practically parallel to the diagonal, $\lambda = L/G = 1$.

Taking into account what has been said, the $x - y$ diagram of the cascade has the form shown in Fig. 3 (in Fig. 4 for the first column). The dotted line shows the operating line in the case of concentration in one column (or in a cascade without flow reduction). Fig. 3 shows that the advantages of cascading are achieved by decreasing the concentration head $\vartheta = y - y_e$, by a loss in the number of transfer units required.

The cascade as a whole, and the operation of the extracting section of the first column are characterized by the degree of isotope extraction from the raw material $D = j/L_0 x_0 = j/j_c$. The operation of the concentrating section of the column may be characterized by the maximum capacity and relative take-off. We consider as maximum the capacity of a concentrating column with an infinite number of plates, when the emergent gas phase has an equilibrium composition $y = y_e$, i.e., $j_0 = Lx_h - Gy_e$. The maximum capacity of a column of a cascade, for which $L = G$, at $x \ll 1$, when $y_e = x/\alpha$, will be *

---

* If $L \neq G$, then $j_0 = Lx_h \dfrac{\alpha\lambda - 1}{\alpha\lambda}$, where $\lambda = L/G'$.

76

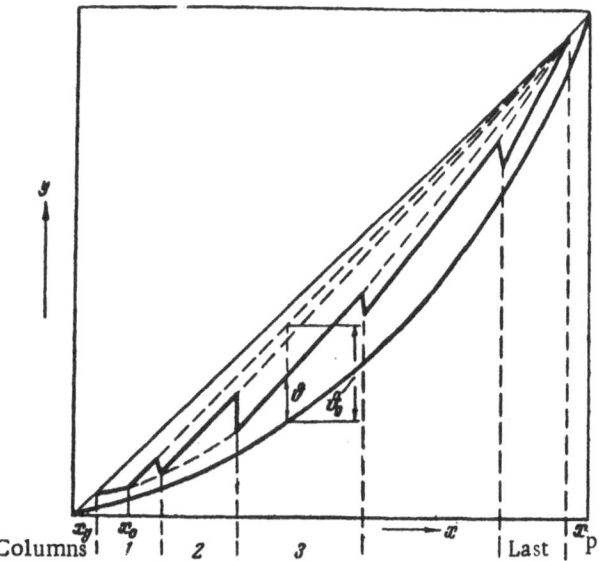

Fig. 3. x – y Diagram of cascade: ϑ_0 – concentration head in a single column or in a cascade without flow reduction; ϑ – the same in a cascade with flow reduction.

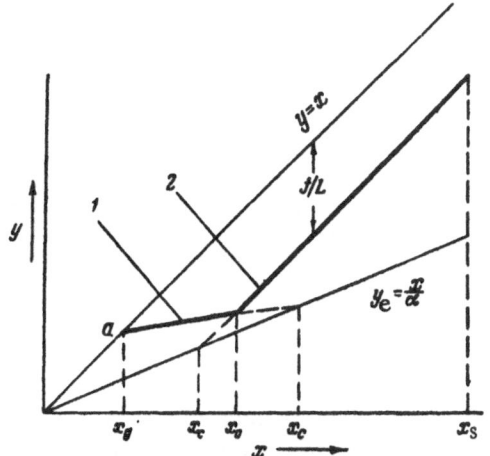

Fig. 4. x – y Diagram of the first column: 1 and 2 – operating lines of the extraction and concentration sections of the column; x_c – concentrations corresponding to the "intersection points."

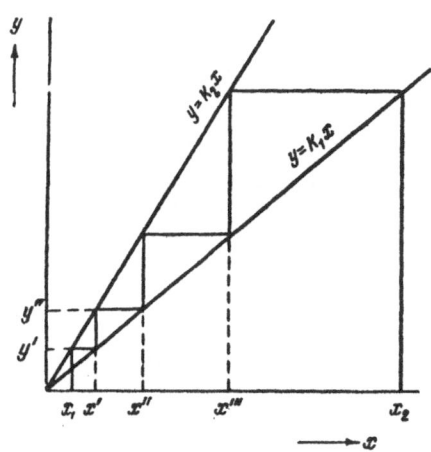

Fig. 5. Determination of the number of steps between the straight lines.

$$j_0 = Lx_h \frac{\alpha - 1}{\alpha} = Lx_h \varepsilon_0. \tag{4}$$

This ratio is also correct at high concentrations, $\underline{x}$ (one should only take into account that $G = L - j$); $\varepsilon_0 = \frac{\alpha - 1}{\alpha}$.

The ratio $\theta = j/j_0$ coincides with the relative take-off $\theta = D/D_{\infty}$ [3]. The material flows in the first column may be found either by calculating the minimum reflux ratio, as is usually done for distillation (i.e., by the conditions in the upper section of the column), or by the maximum capacity (i.e., by the conditions in the lower section of the column). In the first case the minimum reflux ratio is found from Eqs. (1a) and (3a), assuming that $y(x_0) = y_e(x_0) = x_0/\alpha$. If the feed is supplied in liquid form, then

$$R_{\min} + 1 = \frac{\alpha D}{\alpha - 1} = \frac{D}{\epsilon_0} = \frac{j}{j_0(L_0)}. \tag{5}$$

This expression means that the number of times the liquid circulates during fractionation must equal the ratio of the actual capacity to the maximum possible capacity with the use of only one raw material flow L_0. If the column is fed with a vapor–liquid mixture with a gas content κ, then

$$R_{\min} + 1 \approx \frac{D}{\epsilon_0} + \kappa\,(1-D). \tag{5a}$$

Then, taking the actual reflux ratio

$$R_r + 1 = K(R_{\min} + 1), \tag{6}$$

where $K > 1$, and the material flows are found

$$L_1 = L_0(R_r + 1); \quad L_r = L_0 R_r. \tag{7}$$

In the second case, the relative take-off θ_1 is given, and the load on the first column is determined from Eq. (4):

$$L_1 = j/x_0\,\epsilon_0\theta_1. \tag{7a}$$

It is readily seen that the two methods of calculation are equivalent, and

$$K = 1/\theta = j_0/j. \tag{8}$$

The value of K or θ is selected by the responsible engineer, and the choice is usually based on the optimal conditions of energy consumption and cost of columns. Usually $K = 1.1$ to 1.5. The material flows in the subsequent columns are determined by the ratio $L_2 = L_1/\sigma_1$; $L_3 = L_2/\sigma_2$, etc., where σ is the flow reduction coefficient, which is taken as equal to or somewhat less than the calculated one in the corresponding column.

2. Enrichment Produced by Columns Operating at Low Isotope Concentrations ($x < 0.1$).

The degree of separation q_0, produced by a column or any cascade in operation without take-off, is well known [8] ($q_0 = x_s/x_h = \alpha^{N_T} = e^{\epsilon_0 N_P}$, N_T and N_P are the numbers of theoretical plates and transfer units, $\epsilon_0 = (\alpha - 1)/\alpha$). Thus, the purpose of the calculations is to establish the relation between the degree of separation and the capacity. As is shown below, this relation becomes quite simple when we select as operating variables the enrichment $\tilde{q} = q - 1$ (instead of the degree of separation q, which is usually used) and the capacity with respect to pure product j (instead of the amount of product collected, $P = j/x_p$).

To determine the number of theoretical plates N_T within the region of Henry's law, the following property of straight lines may be used: in the case of Fig. 5, $x_2/x_1 = (K_2/K_1)^N$, where N is the number of stages. Transference of the origin of the coordinates to the intersection point of the operating line with the curve of equilibrium x_c, and taking into consideration the fact that $K_2 = \lambda$, $K_1 = 1/\alpha$, gives the relation

$$(x_2 - x_c)/(x_1 - x_c) = (\alpha\lambda)^{N_T} = \tilde{q}_0; \tag{9}$$

$$\frac{x_2}{x_1} - 1 = (\tilde{q}_0 - 1)\left(1 - \frac{x_c}{x_1}\right). \tag{9a}$$

It may be shown that in calculating by the method of transfer units, Eqs. (9)–(9a)* are also obtained, except that

$$\tilde{q}_0 = e^{\frac{\alpha\lambda - 1}{\alpha\lambda} N_P} = e^{\epsilon N_P}. \tag{9b}$$

---

* This follows from the expression for the number of transfer units $N_P = (y_2 - y_1)/\bar{\vartheta}$, where $\bar{\vartheta}$ is the average concentration head, $\bar{\vartheta} = (\vartheta_2 - \vartheta_1)/2.3 \log \vartheta_2/\vartheta_1$. Considering the equations of the operating line $y = \lambda x + c$, and the equilibrium curve $y_e = x/\alpha$, we find $y_2 - y_1 = \lambda(x_2 - x_1)$; $\vartheta_2 - \vartheta_1 = (\lambda - 1/\alpha)(x_2 - x_1)$, so that $N = (1/\epsilon)\ln \vartheta_2/\vartheta_1$. Having noted that $\vartheta_2/\vartheta_1 = (x_2 - x_c)/(x_1 - x_c)$, we obtain (9a) and (9b).

TABLE 1

The Main Relations for Columns Operating at $x \ll 1$ *

Parameters	Extracting section of first column	Concentrating section of first column and all subsequent columns
Abscissa of intersection point of operating line with equilibrium curve	$x_c = \dfrac{\alpha x_0 (1-D)(1-\lambda)}{1-\alpha\lambda} =$ $= x_d (1-\lambda)/\varepsilon_0 \quad (10)$	$x_c = j/L\varepsilon_0 = \theta x_h \quad (14)$
Number of transfer units required	$N_p = \dfrac{2,3}{\varepsilon_0} \lg \dfrac{1-j/j_0}{1-j/j_1^0} \quad (11)$	$N_p = \dfrac{2,3}{\varepsilon_0} \lg \dfrac{q-j/j_0}{1-j/j_0} \quad (15)$
Number of theoretical plates required	$N_T = \dfrac{\lg \dfrac{1-j/j_0}{1-j/j_1^0}}{\lg(1+\varepsilon_0)} \quad (12)$	$N_T = \dfrac{\lg \dfrac{Kq-1}{K-1}}{\lg \alpha} = \dfrac{\lg \dfrac{q-\theta}{1-\theta}}{\lg \alpha} \quad (16)$
Relation between enrichment and take-off		$x_s = x_r q_0 - j(q_0-1)/L\varepsilon_0$ $\bar{q} = \dfrac{K-1}{K}\bar{q}_0 = \bar{q}_0(1-j/j_0) \quad (17)$
Dependence of concentration on height	$x = x_d \left[1 + \dfrac{1}{j_0/j_1^0-1}(1-e^{-\varepsilon_0 \bar{z}}) \right] \quad (13)$	$x = x_h e^{\varepsilon_0 \bar{z}} - j(e^{\varepsilon_0 \bar{z}}-1)/L\varepsilon_0 \quad (18)$

---

*Formulas from (16) and (17), and others, were obtained by the author and V. Kalinin in 1946 using the value of K; Formula (16) was obtained by N. Zhavoronkov and Ia. Zel'venskii [3]; the latter introduced the value λ into Eq. (16).

Simultaneously solving Eqs. (1a), (2), and (3a), one can readily find the coordinates of the intersection points x_c of the corresponding columns and determine the required number of plates or transfer units and the relation between enrichment and take-off.

The results of the calculations are summarized in Table 1.

In Table 1, $x_d = x_0(1-D)$ is the distillate concentration (at the head of the first column); $q_0 = e^{\varepsilon_0 N_P} = \alpha^{N_T}$; $\varepsilon_0 = \dfrac{\alpha-1}{\alpha}$; $j_0 = L_0 x_0$; $\bar{z} = z/h$. (We should note that in such a determination of q_0, the separating effect in the still is not considered; when the latter is considered, $q'_0 = \alpha q_0$); $\varepsilon_c = \dfrac{1}{\lambda}\left(\dfrac{1}{R+1} - \varepsilon_0\right) = \dfrac{\varepsilon_0}{\lambda}\left(\dfrac{j_c}{j_1^0} - 1\right)$; $\lambda = L_r/L_1 = R_r/(R_r+1)$; $j_1^0 = L_1 x_0 \varepsilon_0$.

The distribution of concentration through the column may be calculated using the mass transference equation [7, 9], $dM = Ldx = k\vartheta dF = ks\vartheta dz$, and hence

$$\frac{dx}{dz} = \frac{y - y_p}{\lambda h} = \frac{\vartheta}{\lambda h}, \qquad (19)$$

where $\vartheta = y - y_e$ is the concentration head, and $h = G/kF$ is the height of the transfer unit. The corresponding formulas, obtained by integration Eq. (19), are given in Table 1 and the graph in Fig. 6.

The relation between the degree of separation and take-off (capacity) is especially simple in the case of concentrating columns of the cascade. Eq. (17) may be rewritten in the form

$$\bar{q} = q - 1 = (q_0-1)(1 - j/j_0). \qquad (17a)$$

Thus, the enrichment $q-1$ depends linearly on the take-off (Fig. 7), while the value $q_0 = e^{\varepsilon_0 N_P} = \alpha^{N_T}$ is the degree of separation achieved in working without take-off (separating capacity of column). An increase in take-off leads to a fall in enrichment as the concentration head decreases $\vartheta = y - y_e = \varepsilon_0 x - j/L$; is minimal at the head of the column: $\vartheta_h = (j_0-j)/L$, so that at $j = j_0$, $\vartheta_h = 0$ and, consequently $\bar{q} = 0$. It should be noted, however, that the simple relation (17a) occurs only in a cascade column, when the take-off is small in comparison

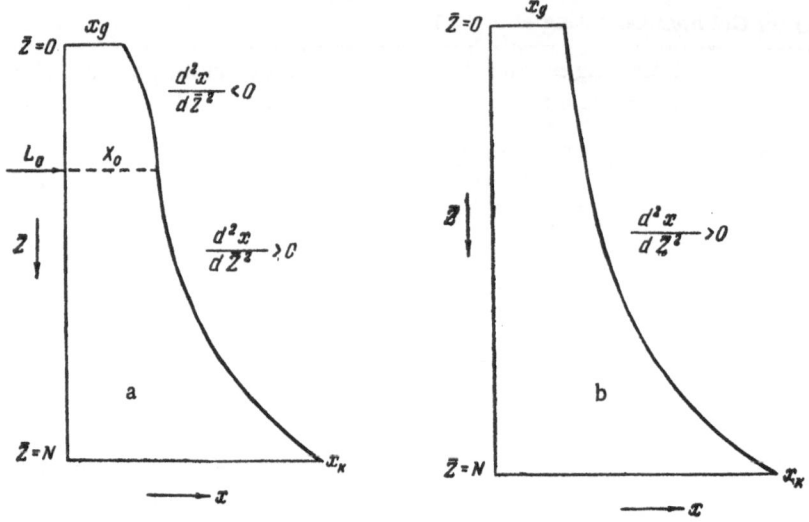

Fig. 6. Concentration distribution through column at x ≪ 1: a) column with fractionation (upper section operating on extraction and lower on concentration); b) concentrating column.

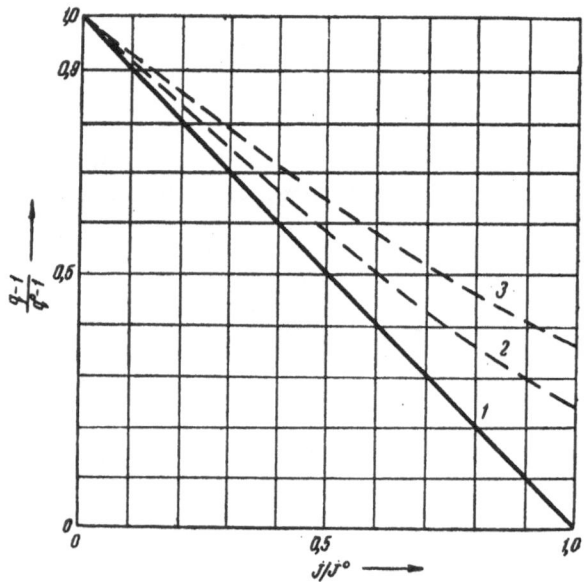

Fig. 7. Dependence of enrichment on capacity: 1) cascade column; 2) and 3) single columns. $q_2^0 = 20$, $q_3^0 = 10$

with the reprocessing flows and does not affect their ratio λ [the operating line remains parallel to the diagonal and, with a change in take-off, is only displaced in accordance with Eq. (3a)]. In the case of a single column,

however, (19) may be integrated or (17a) rewritten, substituting q_0 and j_0 by $\tilde{q}_0 = e^{\frac{\alpha\lambda - 1}{\alpha\lambda} N_P} = q_0^{1+\bar{p}}$; and j_0 by $j_0 = j_0 + j/\alpha q$, which gives

$$\bar{q} = q - 1 \cong (q_0^{1+\bar{p}} - 1)/(1 + \bar{\pi} q_0^{1+\bar{P}}), \tag{17b}$$

where

$$\bar{p} = P/L\epsilon_0 = p/\epsilon_0; \quad q_0 = e^{\epsilon_0 N}; \quad \lambda = L/(L-P) \approx 1 + n; \quad p = P/L.$$

80

Expanding (17b) into a series, and expressing enrichment in terms of capacity $\underline{j}$, we obtain

$$\bar{q} = \bar{q}_0\left(1 - \beta_1 \frac{j}{j_0} + \beta_2 \frac{j^2}{j_0^2} - \cdots\right),\tag{17c}$$

where

$$\beta_1 = 1 - \delta; \quad \beta_2 = \left[1 - \delta\left(1 + \tfrac{1}{2}\ln q_0\right)\right]\Big/q_0; \quad \delta = \ln q_0/(q_0 - 1).$$

Thus, the dependence of enrichment on take-off becomes nonlinear (Fig. 7), and the deviation from (17a) becomes greater the smaller q_0. With an increase in take-off, the enrichment in a normal column decreases less than in a cascade column, which is due to a certain gain in concentration head or an increase in separating capacity of the column with an increase in take-off [due to the increase in the ratio of the flows $\lambda = 1 + p$ and the effective separation coefficient $\alpha\lambda = \alpha(1 + p)$]. In addition, j_0 also increases (to j_c). However, the deviation from (17a) is not very great (and not appreciable for highly efficient columns, when $q_0 > 20$). This is explained by the fact that the determining value has a minimum concentration head (in the head of the column), which is the same in single and in "cascade" columns.

Another example of the effect of λ is the operation of a cascade column with supplementary flows. If the still of the column is supplied with a liquid or gas in an amount ΔL and with concentration x_e, then the ratio of the flows will be $\lambda = 1/(1 + \Delta L/L) < 1$. Considering (9), instead of (17a) we will obtain

$$q - 1 = (\tilde{q}_0 - 1)\left(1 - \frac{j - j_e}{j_0}\right),\tag{17d}$$

where

$$j_e = \Delta L \cdot x_e; \quad \tilde{q}_0 \approx q_0 e^{-\frac{\Delta L}{L}N}; \quad \tilde{j}_0 = j_0\left(1 - \frac{\Delta L}{L\varepsilon_0}\right)$$

In particular, if $x_e = 0$, then the enrichment produced by the column will fall, not so much due to dilution in the still as to a change in the ratio of the flows and a decrease in the separating capacity of the column $\tilde{q}_0$.

The relation $q(j)$ is more complicated for the first column, which combines concentration with extraction. This relation is essentially related to the position of the point at which the feed is supplied; the optimal height for the imput point (with regard to this, one is usually limited by qualitative considerations [10]) may be calculated, assuming that complete mixing of the reflux and the raw material occurs on the feed plate and the concentration of the liquid phase changes abruptly (from x_n to x_h), while the concentration of the gas phase $\underline{y}$ remains unchanged (x_n is the reflux concentration and x_h is the concentration after mixing, $L_1x_h = L_Tx_n + L_0x_0$).

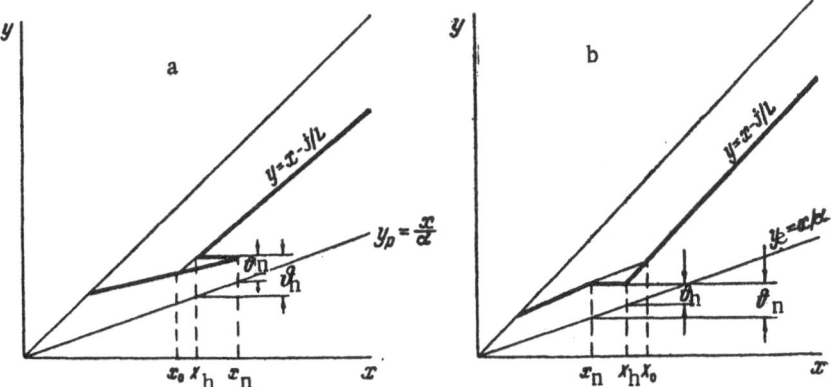

Fig. 8. Jumps and fall in concentration head with nonoptimal feed imput in the first column: a) too low an imput; b) too high an imput.

Then, calculating x_n by Formula (13), we may find x_h, and by Eq. (17), the concentration in the still of the column x_s. From the optimal condition $dx_s/d\bar{z} = 0$, we find

$$\left(\frac{dx_h}{d\bar{z}}\right)_{opt} = \varepsilon_0 x_h - j/L_1 = \vartheta_h.\tag{19a}$$

Using Eq. (19), one can show that condition (19a) indicates equality of concentration heads before and after mixing: $\vartheta_n = \vartheta_h$. In other words, the optimal imput corresponds to the intersection point of the operating lines in the two sections of the column; with deviations from the optimal, ϑ changes abruptly and part of the plates operate in a zone with decreased concentration head (Fig. 8) and enrichment falls. Incidentally, the accuracy requirement in finding the optimal height is not too strict. With small $\Delta\bar{z} = \bar{z} - \bar{z}_{opt}$ $q_{opt} - q = \beta(\Delta\bar{z})^2$, where $\beta = q_1^0 \varepsilon_0^2 \dfrac{j_c(1-\theta)}{j_1^0} \dfrac{}{2}$; at $\alpha - 1 = 0.03$, $j_c/j_1^0 = 2$, $q_1^0 = 10$, $\theta = 0.5$, we obtain $\Delta q = 4 \cdot 10^{-3} \Delta\bar{z}^2$.

If one tries to maintain the optimal conditions for feed imput, an increase in capacity must be accompanied by an increase in the number of plates in the extracting section — $\bar{z}_{ext}$, i.e., by lowering the imput point. Correspondingly, the number of concentrating plates decreases, $N_1 = N - \bar{z}_{ext}$, q_0 and likewise with the enrichment q_1. In this case, Eq. (17a) could be applied, by the magnitude $q_1^0 = e^{\varepsilon_0 (N - \bar{z} ext)}$ would depend on capacity; substituting $\bar{z}$ in (11), we find

$$q_1^0 = q_1^{00} \left(\frac{1 - j/j_1^0}{1 - j/j_c} \right)^{\frac{\lambda}{(j_1^0/j_0) - 1}}, \tag{20}$$

where $q_1^{00} = e^{\varepsilon_0 N}$ is the total separating capacity of the first column. It follows from (20) and (17a) that in this case the relation $q(j)$ will be nonlinear (Fig. 9). However, if the height of the imput is not varied, then it will

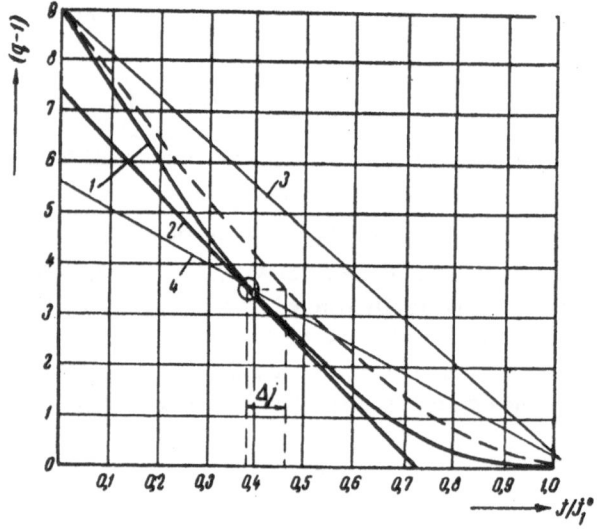

Fig. 9. The dependence of enrichment on capacity for the first cascade column with a moveable optimal feed imput (curve 1) and with constant height of imput (curve 2). The analogous relations for a concentrating column, with a separating capacity of q_1^{00} (curve 3) and q_1^0 calculated (curve 4), are given for comparison. The dotted line shows a variation of curve 1 with the feed increased by 50%. $\odot$ is the rated state.

not be optimal for all the take-offs except one [for which the number of plates in the extraction zone will satisfy Eq. (11)]; this take-off we will call rated. The nonoptimal state will lead to a decrease in enrichment (Fig. 9, curve 2), but the characteristic of $q(j)$ will be linear, touching the optimal curve at the calculated take-off:

$$q = \frac{q_1^0}{1 - D_e} \left\{ 1 - \frac{j}{j_1^0} \times \left[\frac{q_1^0 - 1}{q_1^0} (1 - D_e) + \frac{j_1^0}{j_c} \right] \right\}, \tag{21}$$

where D_e is the degree of extraction under rated conditions.

3. Columns Operating at High Isotope Concentrations

In this case the required number of transfer units may be found using the known equation:

$$N = \int \frac{dy}{\vartheta} = \lambda \int_{x_h}^{x_s} \frac{dx}{y - y_p},$$

where y is determined by Eq. (3) and y_e by Eq. (1). Integration gives:

$$\left(\frac{x_s - x_{c1}}{x_h - x_{c1}}\right) \cdot \left(\frac{x_{c2} - x_h}{x_{c2} - x_s}\right)^{1/\bar{\alpha}} = e^{\frac{\bar{\alpha}-1}{\bar{\alpha}} N} = \tilde{q}_0, \qquad (22)$$

where x_{c1} and x_{c2} are the coordinates of the intersection points of the operating line with the equilibrium curve (Fig. 10), $\bar{\alpha} = [\alpha - (\alpha - 1)x_{c1}]/[\alpha - (\alpha - 1)x_{c2}]$.

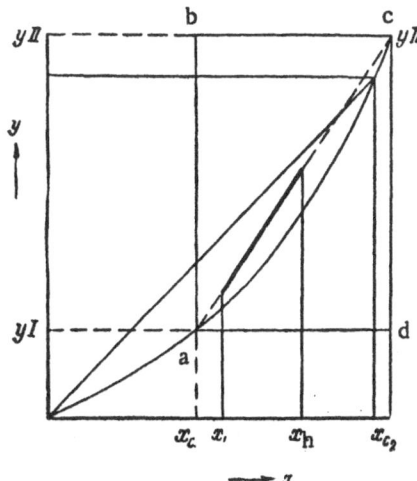

Fig. 10. Intersection points of the operating line with the equilibrium curve at high concentrations of the component being extracted.

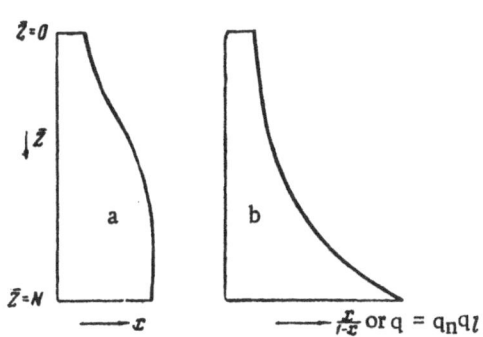

Fig. 11. Distribution of concentration with respect to height (in the case of high concentrations of the components being extracted): a) absolute concentration; b) relative concentration.

An analogous result, but by a more complicated method, may be obtained for the number of theoretical plates (by transforming the rectangle abcd into a square and the operating line into a diagonal [11] and by applying Fenske's formula [8]. In this case, $\tilde{q}_0 = \bar{\alpha}^N$; thus $\bar{\alpha}$ is the separation coefficient in the transformed system of coordinates. Generally, x_{c1} and x_{c2} may be found from the condition that $y = y_e$ as the roots of the quadratic equation

$$x^2 - x\left[1 + \frac{j}{\varepsilon_0 L}\left(1 + \frac{1}{\alpha}\frac{1-x_p}{x_p}\right)\right] + \frac{j}{\varepsilon_0 L} = 0. \quad (23)$$

If, however, the cascade produces a highly concentrated product, $1 - x_p \approx 0$, then the solution acquires a very simple form:

$$\begin{aligned} x_{c1} &= x_c = j/L\varepsilon_0 = \theta x_h \\ x_{c2} &= 1, \end{aligned} \qquad (24)$$

and x_{c1} coincides with the value found previously [Eq. (14)]. In this case a formula may be obtained from (22), which would be a generalized expression of (15)-(17):

$$\frac{q_n - \theta}{1 - \theta} q_l^{1/\bar{\alpha}} = \tilde{q}_0 = e^{\varepsilon_0 N (1 - x_c)}, \qquad (25)$$

$$N \approx \frac{1}{\varepsilon_0 (1 - x_c)} \ln \frac{q_n - \theta}{1 - \theta} q_l, \qquad (26)$$

$$q - 1 = \bar{q} = (\tilde{q}_0 - 1)(1 - \theta)/(1 - x_c), \qquad (27)$$

where $q_n = x_s/x_h$ and $q_l = (1 - x_h)/(1 - x_s)$ are the enrichments with respect to the less volatile and the more readily volatile components, $q = q_n q_l$ (at $x_h < 0.1$ one may assume that $x_c = 0$).

In comparing expressions (15)-(17a) and (25)-(27), one may conclude that the difference in the case of high concentrations consists mainly of the necessity for considering the enrichment with respect to the volatile component. The other conclusion concerns cases of low concentration: one may see from (26) that for a cascade column with $x \ll 1$ one should neglect the slight increase in the effective separation coefficient $\alpha_{ef} = \alpha\lambda$ connected with the take-off (due to which the ratio of flows λ exceeds unity: $\lambda = 1 + p > 1$). Actually, it follows from (26) that $\varepsilon_{ef} \approx \alpha_{ef} - 1 = (\alpha - 1)(1 - x_c) < (\alpha - 1)$, while $\alpha\lambda - 1 > (\alpha - 1)$. This is explained by the fact that the loss in concentration head due to distortion of the equilibrium line exceeds the gain from the deviation of λ from unity. We therefore did not introduce the value λ (we used $\lambda = 1$) in Formulas (15)-(17).

Fig. 11 shows the distribution with respect to column height of the absolute concentration $\underline{x}$ and the relative concentration $X = x/(1-x)$; the latter hardly differs from the distribution at low concentrations (cf. Figs. 6 and 11). We should also note that when $\tilde{q}_0 = f(j)$ then the relation $q(j)$ becomes nonlinear.

The expressions (24) and (25) at $x_p \geq 0.9$ increase the calculated number of theoretical plates inappreciably (by less than 1%); if necessary, a correction for the deviation of x_e from unity may be introduced into (24): $x_{c_1} \approx \theta x_h (1 - \delta)$; $x_{c_2} \approx 1 + \delta$, where $\delta = \theta x_h (1 - x_p)/(1 - \theta x_h) \alpha x_p$.

If the concentration of the product $x_p < 0.8$, the number of theoretical plates in the last column should be determined from (22) and x_{c_1} and x_{c_2} calculated from (24). Then

$$N \approx \frac{\overline{\alpha}}{\overline{\alpha} - 1} \, 2,3 \lg \frac{x_s - x_{c_1}}{x_h - x_{c_1}} \cdot \frac{x_{c_2} - x_h}{x_{c_2} - x_s}.$$

4. The Cascade as a Whole.

The relation between enrichment and capacity has the simplest form for a cascade of concentrating columns operating in the range of low concentrations [4]. In this case, by rewriting Eq. (17a) for enrichment in column "e" as

$$(x_s)_e = q_e^0 (x_s)_{e-1} - (q_e^0 - 1) j/L_e e; \qquad q_e = q_e^0 \left(1 - \frac{q_e^0 - 1}{q_e^0} \frac{j}{i_e^0}\right), \tag{17d}$$

successively combining Eqs. (17d) for the separate columns, and noting that $j_{e+1}^0 = j_e^0 q_e / \sigma_e = j_1^0 q_1 q_2 \dots$, $q_e / \sigma_1 \sigma_2 \dots \sigma_e$, we find enrichment Q, produced by the cascade:

$$Q_n = Q_n^0 (1 - jR/j_1^0) = Q_n^0 (1 - j/j_0), \tag{28}$$

where $Q_n = q_1 q_2 \dots q_n = \frac{x_n}{x_0}$; $Q_n^0 = q_1^0 q_2^0 \dots q_n^0$ is the separating capacity of the cascade; j_1^0 is the maximum capacity of the first column; $j_0 = \frac{j_1^0}{R}$ is the maximum capacity of the cascade, and R is the resistance coefficient of the cascade; $R = r_1 + r_2 + \dots + r_n$, the resistance coefficient of the column:

$$r_1 = \frac{q_1^0 - 1}{q_1^0} = \frac{\overline{q}_1^0}{q_1^0}; \qquad r_2 = \frac{\sigma_1}{q_1^0} \frac{\overline{q}_2^0}{q_2^0}; \qquad r_n = \frac{L_1 \overline{q}_n^0}{L_n Q_n^0} = \frac{\sigma_1 \sigma_2 \dots \sigma_{n-1} \, \overline{q}_n^0}{q_1^0 q_2^0 \dots q_{n-1}^0 q_n^0}. \tag{29}$$

Thus, as for a separate column, the enrichment depends linearly on take-off for the cascade examined (Fig. 12); the magnitude $j_0 = j_1/R$ is practically both the maximum and the actual capacity of the cascade, as usually $Q \ll Q_0$ ($\gamma = Q/Q_0 \sim 10^{-2}$), while $j = j_0 (1 - \gamma)$. The values R and $\underline{r}$ we call resistances as they determine the capacity of the cascade ($j_0 = j_1^0/R$) and of the columns, with r_i being added as with electrical resistances. With columns connected in series $R = \sum_{i=1}^{n} r_i$ with parallel connections we add $1/r$; thus, for a cascade with a by-pass (Fig. 13)

$$R \approx r_1 + \frac{1}{1/(r_2 + r_3) + 1/(r_2' + r_3' + r_4')} + r_4 + r_5$$

(r' is the resistance of the by-pass columns).

With these conditions, the value of the resistance of each column r_k is proportional to the degree of flow reduction $\sigma_1 \sigma_2 \dots \sigma_{k-1} = L_1/L_k$ and inversely proportional to enrichment Q_{k-1}^0 (the capacity or "conductivity" of the column, $1/r_k$ is proportional to flow L_k and enrichment Q_{k-1}^0).* For an ideal cascade, whose columns have an equal number of theoretical plates N and a relative take-off θ, as well as a degree of flow reduction which coincides with enrichment ($\sigma_k = q_k$), the resistance coefficients decrease with an increase in the number

_____
*If the potentials $\varphi_n = Q_n /Q_n^0$ and the currents $I = j/ j_1^0$ are introduced, the analogy to electricity becomes complete: $\varphi_{n-1} - \varphi_n = I_n r_n$. In a nonstationary process, the holding capacities of the columns act as condensers $C_n \, d\varphi_n /dt = I_n - I_{n+1}$; $C_n = E_n Q_n^0 x_0 /j_1^0$; E_n is the holding capacity of the still of the n-th column. We should note that the resistances of the columns include the decrease in concentration head due to the steps shown in Fig. 3.

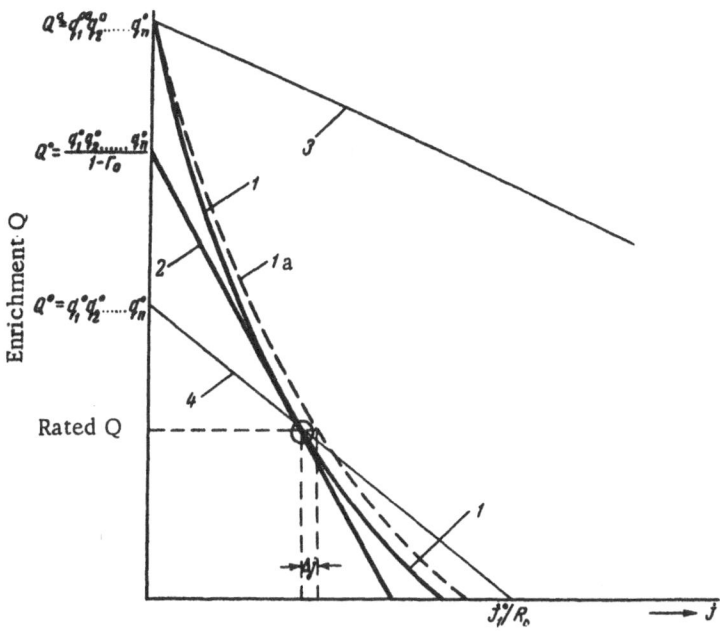

Fig. 12. Relation of enrichment produced by a cascade of columns to capacity: curves 1 and 2 – with moveable and immoveable feed imput in the first column (1a – curve 1 with increased feed, Δj is the corresponding gain in capacity); 3 and 4 – cascade of concentrating columns with a separating capacity for the first column q_1^{00} and q_1^0 rated; $\odot$ – rated state.

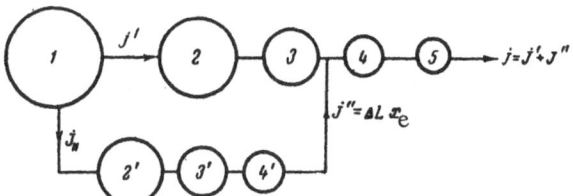

Fig. 13. Cascade with by-pass.

of columns: $r_{k+1}/r_k \approx (1 - \theta)$, so that $r_k \approx (1 - \theta)^k$, while $R = 1/\theta = K.$[*] Thus, capacity is limited by the first columns (at $\theta = 0.5$ by the first four: $r_1 \approx 1$, $r_2 \approx 0.5$, $r_3 \approx 0.25$, $r_4 \approx 0.12$, $r_5 \approx 0.06$; and at $\theta = 0.8$ by two: $r_1 \approx 1$, $r_2 \approx 0.2$, $r_3 \approx 0.04$). Approximately the same relations exist for any correctly planned cascade for which $\sigma_k \leq q_k$. If, however, the enrichment produced by a column is less than the degree of flow reduction, then the following column does not pass the required flow of isotope and becomes a narrow section in the cascade; correspondingly, its "resistance" coefficient increases (for example, in the case of a plant for water distillation in the USA [12] – r_2 and, probably, r_3).

If the first column of the cascade operates with fractionation, then the characteristic of $Q(j)$ of the cascade is analogous to the characteristic of the first column (Fig. 12).

If the last columns (for example, the 4th and 5th) operate at high concentrations of the component being extracted, then Eq. (28) is applicable; with this, Qn includes enrichment with respect to the low-volatility

---

[*]Here $K = (R_{ref} + 1)/(R_{min} + 1)$; see Eq. (6).

component: $Q_n = Q_n Q_1 = x_p / x_0 (1 - x_p)$. In the last columns, q_0 should be replaced by $\overline{\overline{q}}_0 = (q_0^{1-x_0}) / (1 - x_0)$,

and $r_4 = \dfrac{\sigma_1 \sigma_2 \sigma_3}{q_1^0 q_2^0 q_3^0} \dfrac{\overline{\overline{q}}_4^0 - 1}{\overline{\overline{q}}_4^0}$; $r_5 = \dfrac{\sigma_1 \sigma_2 \sigma_3 \sigma_4 q_{1_4} (\overline{\overline{q}}_5^0 - 1)}{q_1^0 q_2^0 q_3^0 \overline{\overline{q}}_4^0 \overline{\overline{q}}_5^0}$; correspondingly, the relation Q(j) of the cascade be-

comes nonlinear.

In the case where there is loss of product in the stills of the columns and in their connections, the resistance of the cascade increases while the enrichment achieved falls:

$$Q = Q_0 \left[1 - \sum_{l=1} r_l j^{(l)}) / j_1^0 \right] = Q_0 (1 - j\widetilde{R}/j_1^0), \tag{28a}$$

where j_l is the actual take-off from the l-th column (allowing for losses),

$$\widetilde{R} = \Sigma r_l \, j_l / j = \Sigma r_l (1 + \delta_l) = R + \sum_{l=1}^{n} r_l \delta_l; \quad \delta_l = j_l / j - 1 - \text{coefficient of loss.}$$

If the losses in the i-th column (with respect to 100% product, in unit time) are Δi, then $j_1 = j + \sum_{i=1}^{n} \Delta_i$,

$j_l = j + \sum_{i=l}^{n} \Delta_i$, and hence $\delta_1 = \left(\sum_{i=1}^{n} \Delta_i \right) / j$, $\delta_l = \left(\sum_{i=l}^{n} \Delta_i \right) / j$, $\delta_n = \Delta_n / j$.

In conclusion, we should note that Eq. (28) is also applicable to a cascade with elements with a single-stage separating action and flow reduction [if the flows are not reduced, then such a cascade is equivalent to one column and Eqs. (17a) or (17b)] would be accurate, for example, for a cascade of electrolyzers. In this, the separating capacity of columns q^0, in Formulas (28) and (29), would be replaced by the separating capacity of an element $\alpha_{ef} = \alpha \lambda$, where $\lambda_i = L_i / (L_i - P)$. We obtain $Q_n = \alpha^n \left(1 - R \dfrac{j}{j_0} \right)$, where $j_0 = L_0 x_0$; $R = 1 +$

$+ \dfrac{\sigma_1}{\alpha \lambda_0} + \dfrac{\sigma_1 \sigma_2}{\alpha^2 \lambda_0 \lambda_1} + \cdots$ so that $Q_n \approx Q_n^0 (1 - j / j_{lim})$, the $x - y$ diagram of such a cascade is analogous to the one shown in Fig. 3, except that only the lower points of the "teeth" are achieved practically.

5. Columns and Cascades Operating by the Two-temperature Method (Using Isotope Desorption or Absorption for Phase Transition).

As has been indicated [4], phase transition is possible by the use of isotopic "absorption" processes. For example, if in an exchange an isotope is concentrated in a liquid phase (distribution coefficient a = y/x < 1), then to induce flow circulation, instead of using evaporation, one may use a "desorption" column, operating at a higher temperature and ensuring sufficiently complete removal of the isotope being concentrated from the emergent flow (on the contrary, at a > 1, the concentration would occur in the desorption column and the absorption one would act as condenser. The plan and $x - y$ diagram for such a process are shown in Fig. 14 (in the case of extraction of the isotope from a liquid phase flow L); the isotope accumulates in the still of the "cold" column and in the upper section (head) of the "hot" column. It was noted [4] that in this case the position of the operating lines cannot be found directly from the material balance; a simultaneous solution of the balance equations and the equations relating concentration to the separating capacity of the columns is required. If the columns operate at low concentrations of the component being extracted, then the equations of the operating lines and the equilibrium curves will be

$$y_{e1} = a_1 x; \quad y_{e2} = a_2 x; \quad y_1 = \lambda_1 x + C_1; \quad y_2 = \lambda_2 x + C_2, \tag{30}$$

where a_1 and a_2 are the distribution coefficients of the isotope at the two temperatures, $\lambda = L/G$ is the ratio of the liquid and gas flows, and C_1 and C_2 are constants, unknown as yet. As can be seen from Fig. 14, when there is take-off $\lambda_1 \neq \lambda_2$, $C_1 \neq C_2$. Fig. 14 also shows that the compositions of the gas, corresponding to the starting liquid ($x = x_0$) and the impoverished one [$x = x_0 (1 - D)$] are the same.

Assuming that $y_1(x_0) = y_2[x_0 (1 - D)]$ and denoting $\lambda_1 - \lambda_2 = P/G = \Delta\lambda$, we find the relation between the constants C_1 and C_2:

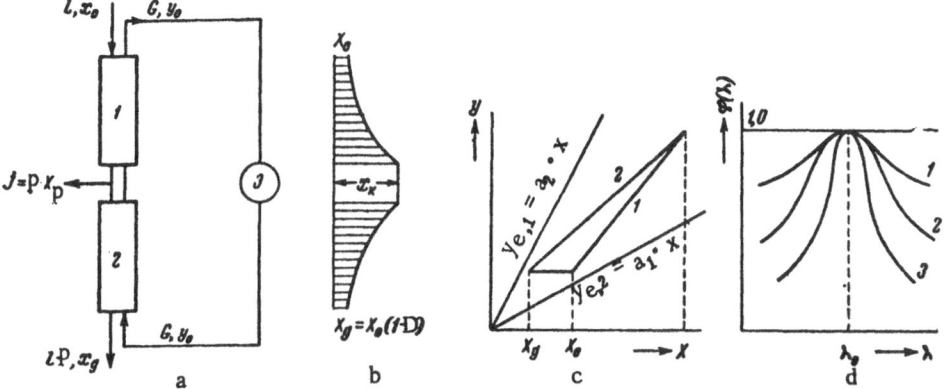

Fig. 14. Concentration by the two-temperature method: a) plan; 1 and 2 — columns operating at low and at high temperatures; 3 — gas circulator, $j = Px_p = Lx_0D$ — capacity of the whole system, x_0 and $x_d = x_0(1-D)$ — isotope concentration in the incoming and outgoing liquids; b) concentration distribution of the liquid phase through the height of the column; c) $x-y$ diagram, $a_1 < a_2 < 1$; 1 and 2 — the operating lines of the "cold" and the "hot" columns; d) the relation of enrichment to the flow ratio λ at $N_1 = N_2$; curves 1, 2, and 3 — for columns with $q_1^0 < q_2^0 < q_3^0$.

$$C_1 - C_2 = (\lambda D - \Delta\lambda)x_0. \tag{31}$$

Let us apply the method of straight lines [Eq. (9)] to determine the enrichment in each column. We obtain

$$\frac{x_2 - x_{c2}}{x_0(1-D) - x_{c2}} = \widetilde{q}_2^0; \qquad \frac{x_2 - x_{c_1}}{x_0 - x_{c1}} = \widehat{q}_1^0, \tag{32}$$

where

$$\overline{q}_2^0 = (a_2/\lambda)^{N_{T_2}} = e^{\frac{a_2 - \lambda_2}{\lambda_2}N_{\Pi_2}}; \quad \widetilde{q}_1^0 = (\lambda/a_1)^{N_{T1}} = e^{\frac{\lambda_1 - a_1}{\lambda_1}N_{P_1}},$$

$N_{p1,2}$ and $N_{T1,2}$ is the number of transfer units or theoretical plates in the first and second columns. Then, from (30) and (31) we find the relation between C_1, C_2, x_{C1} and x_{C2}:

$$C_1 - C_2 = x_{c2}(a_2 - \lambda_2) + x_{c1}(\lambda_1 - a_1) = (\lambda D - \Delta\lambda)x_0. \tag{33}$$

Simultaneously solving Eqs. (32) and (33), we find that without take-off, when $D = 0$, $\lambda_1 = \lambda_2 = \lambda$; $C_1 = C_2$, the enrichment appears as

$$\overline{q} = \frac{x_s - x_0}{x_0} = \frac{(\widetilde{q}_1^0 - 1)(\widetilde{q}_2^0 - 1)(a_2 - a_1)}{(a_2 - \lambda)(\widetilde{q}_1^0 - 1) + (\lambda - a_1)(\widetilde{q}_2^0 - 1)}. \tag{34}$$

Analysis of Formula (34) shows that enrichment with two-temperature concentration depends essentially on the ratio of the phase flows $\lambda = L/G$ and reaches the optimal value when $\widetilde{q}_1^0 = \widetilde{q}_2^0 = \widetilde{q}_3^0$. Thus,

$$q_{opt} = q_0, \tag{34a}$$

while the optimal ratio of the flows (in transfer units) appears as

$$\lambda_0 = \frac{a_1N_1 + a_2N_2}{N_1 + N_2} \tag{35}$$

[using the plate method $\lambda_0 = (a_1^{N_1}a_2^{N_2})^{1/(N_1+N_2)}$]. Thus, λ_0 approaches the distribution coefficient of the column that has the greatest number of transfer units (at $N_1 \to \infty$ $\lambda_0 \to a_1$, at $N_2 \to \infty$ $\lambda_0 \to a_2$). If the two columns are the same, $N_1 = N_2$, then $\lambda_0 = (a_1 + a_2)/2$ (using the plate method, we obtain $\lambda_0 = \sqrt{a_1a_2}$). The optimal separating capacity of the columns would appear as

$$(q^0)_{opt} = e^{\frac{\beta-1}{\gamma\beta+1}N_2}, \tag{35a}$$

where $\beta = a_1/a_2$; $\gamma = N_1/N_2$. If $N_1 = N_2$, then $q^0 = e^{\frac{\beta-1}{\beta+1}N}$ [4] (in this case, using the plate method gives $q_0 = \beta^{N/2}$). In this way the two columns have approximately the same separating effect (where $N_1 = N_2 = N$) as one distillation column with N plates and a separation coefficient of $\approx \sqrt{\beta}$.

Taking into account (34a), expression (34) may be rewritten for the enrichment in a two-column system with $j = 0$ as

$$(q - 1)_{j=0} = (q^0_{opt} - 1)\,\psi(\lambda), \tag{34b}$$

where $\psi(\lambda)$ is a function with a maximum, whose form is shown in Fig. 14c. The maximum is sharper the higher the separating capacity of the column; at $N_1 = N_2$ and $q_0 > 10$, with some approximation $\psi(\lambda) = 1/\mathrm{ch}N(\lambda/\lambda_0 - 1)$. If $N_1 \neq N_2$, then the curve $\psi(\lambda) = f(\lambda)$ is unsymmetrical.

Figure 14 also explains the reason for the appearance of maxima. The decrease in q at $\lambda \neq \lambda_0$ is related to the decrease in the concentration head (especially at the ends of the column) when the operating line turns into a narrow "fork" between the two equilibrium curves [4] (remembering that λ is the slope of the operating line).

If the columns operate with take-off, then in solving the system of equations (32)-(33), and neglecting the difference between λ_1 and λ_2 and certain correction terms (i.e., actually solving the problem for a column operating in a cascade when $\lambda_1 = \lambda_2 = \lambda$),* we obtain an expression which differs from the one obtained for a distillation column (17a) only in the correction $\psi(\lambda)$

$$q - 1 = (q^0_{opt} - 1)\,\psi(\lambda)\left(1 - \frac{\beta}{\beta - 1}D\right) = (q_0 - 1)(1 - j/j_0)\,\psi(\lambda), \tag{36}$$

since $D\beta/(\beta - 1) = Lx_0D/Lx_0(\beta - 1)/\beta$; $Lx_0D = j$; $Lx_0(\beta - 1)/\beta \approx j_0$.

The material flows in the system (liquid load L) may be found with respect to the maximum capacity [Formula (4)], using the value β as the separation coefficient and selecting the relative take-off θ in the same way as for distillation: $L = j/\frac{\beta - 1}{\beta}x_0\theta$. The load of the gas phase will be $G = L/\lambda_0$.

The number of transfer units of the "hot" column will appear as

$$N_2 = \frac{\beta + 1}{\gamma\beta - 1}\ln\frac{q - j/j_0}{1 - j/j_0},$$

and for the "cold" one as $N_1 = N_2/\gamma$. With an equal relative take-off from the two columns, the optimal value (with the condition that $N_1 + N_2 = \min$) of $\gamma = \sqrt{\beta}$.

The analogy with distillation may also be used in a cascade of two-temperature columns, whose plan and $x - y$ diagram are shown in Fig. 15. In this case, as $P \ll L$, as in the distillation cascade, the slope of the operating lines λ may be considered as independent of take-off and this leads to expression (36). Transposition of these expressions for each step to appear as (17g), which is convenient for calculating the cascade as a whole, is not difficult. We obtain

$$q_l = \widetilde{\widetilde{q}}{}^0_l\left(1 - \frac{\widetilde{\widetilde{q}}{}^0_l - 1}{\widetilde{\widetilde{q}}{}^0_l}\frac{j}{j^0_l}\right), \tag{36a}$$

where $\widetilde{\widetilde{q}}{}^0_l = 1 + (\widetilde{q}{}^0_l - 1)\psi(\lambda)$; $j^0_l = Lx_{hl}\frac{\beta - 1}{\beta}$ is the maximum capacity of the l-th column.

Combining Eq. (36a) with the calculated relation between the maximum capacities of the columns, we obtain Formula (28) for the cascade as a whole; in calculating the cascade "resistance" R, the values of q_0 are replaced by $\widetilde{\widetilde{q}}_0$.

---

*If these simplifications are not made, then the enrichment must be expressed not in terms of capacity j, but in terms of take-off P. This gives a formula analogous to the one given for a single distillation column (17b).

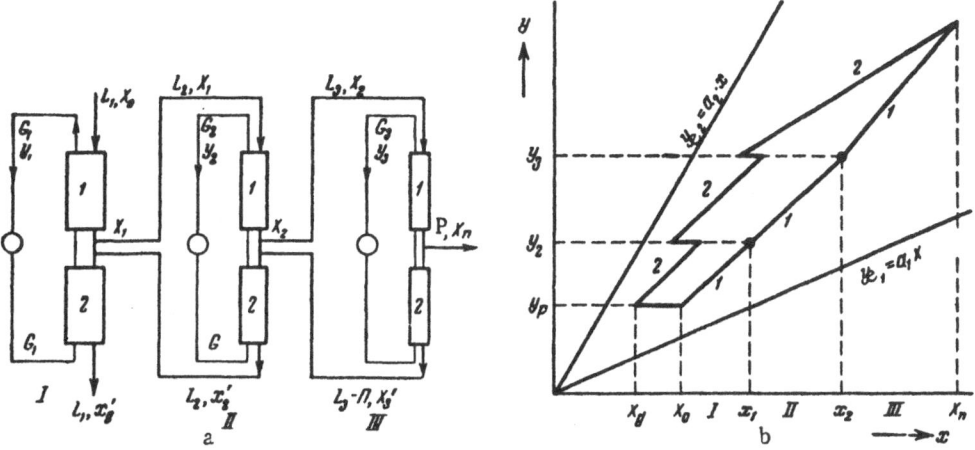

Fig. 15. Cascade of two-temperature columns: a) plan; b) x − y diagram.

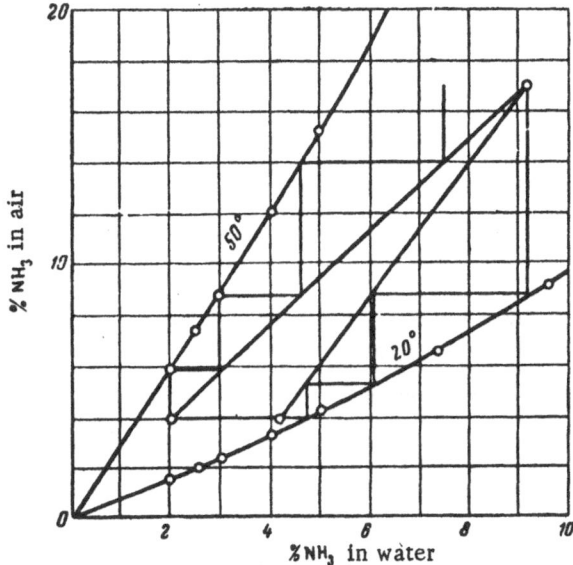

Fig. 16. Two-temperature enrichment of ammonia in water.

In conclusion let us note the following. In producing isotopes by concentration distillation, a method is used that has been known for a long time in chemical technology. On the contrary, the two-temperature method may be used not only for separating isotopes but also for absorption concentration of normal mixtures. Thus, the author and V. F. Kalinin showed, in 1946, that the method could be used for the concentration of ammonia in water (with air circulation, Fig. 16), and they put forward and confirmed Eq. (36). Approximately the same results were obtained by S. Kalinina and N. Kuz'minykh [13] in concentrating sulfur dioxide.

II. Nonstationary Process in Single Columns and in Cascade

1. The solutions of the problem of the rate at which columns reach equilibrium are either extremely complex [2, 14] or are based on various assumptions on the concentration distribution in the columns [1, 6]. Nonetheless, a simple solution without any assumptions may be attained, starting only from the general equation of nonstationary mass transfer [5].

The latter differs from Eq. (19) in the term which accounts for the accumulation of the component in the liquid, retained by the packing or the plates, and appears as:

$$\frac{\partial x}{\partial z} + \omega\frac{\partial x}{\partial t} = \frac{\vartheta}{\lambda h} = \frac{y - y_{\text{p}}}{\lambda h}, \tag{37}$$

where $\omega = \overline{\Omega}/L$, $\overline{\Omega} = \Omega/H$ is the holding capacity of one meter of packing; Ω is the total holding capacity of the packing or plates; H is the height of the column; L is the liquid load.

The accumulation of the component in the column below a point $\underline{z}$ at the moment $\underline{t}$ will be

$$M(z,t) = \overline{\Omega}\int_z^H (x - x_0)\,dz + E_{\text{s}}\,(x_{\text{s}} - x_0) = M_{\Omega}(z,t) + M_{\text{s}}\,(t) \tag{38}$$

where $M_{\Omega}(z, t)$ is the accumulation of the component in the liquid, retained by the packing; $M_{\text{s}} = E_{\text{s}}(x_{\text{s}} - x_0)$ is the same in the still; E_{s} is the holding capacity of the column's still.

The total accumulation in the column $M = M_{\Omega}(0, t) + M_{\text{s}}$ will be denoted as a magnitude without an index.

On reaching equilibrium, the total accumulation in the column will be

$$M_{\text{p}} = M_{\Omega}(0,\infty) + M_{\text{s}}(\infty) = \overline{\Omega}\int_z^H (x_{\text{p}} - x_0)\,dz + E_{\text{s}}\,(x_{\text{es}} - x_0) =$$

$$= \left[\Omega\left(\frac{q_0 - 1}{\ln q_0} - 1\right) + E_{\text{s}}\,(q_0 - 1)\right]x_0 = \left[\left(\frac{\Omega}{\varepsilon_0 N} + E_{\text{s}}\right)(e^{\varepsilon_s N} - 1) - \Omega\right]x_0 \tag{38a}$$

where x_e is the equilibrium concentration $x_e = x_0 e^{\varepsilon_0 \overline{z}}$; $q_0 = e^{\varepsilon NP}$.

In one unit of time

$$j(z,t) = \frac{\partial M(z,t)}{\partial t} = \overline{\Omega}\int_z^H \frac{\partial x}{\partial t}\,dz + E_{\text{s}}\frac{dx_{\text{s}}}{dt} = j_{\Omega}(z,t) + j_{\text{s}}(t), \tag{38b}$$

will be extracted below point $\underline{z}$, while in the whole column $j = j_{\Omega}(0, t) + j_{\text{s}}$ will be extracted (the magnitude $\underline{j}$ is sometimes called the transport [2] or transfer [1] of the isotope).

It follows from the material balance of the column $Lx - Gy = j(z, t)$ that

$$y = x - j(z, t)/L, \tag{38c}$$

i.e., the equation of the operating line has the same form as in the stationary state [Eq. (3a)], but the outer take-off $j_0 = Px_{\text{s}}$ is substituted by the inner $\underline{j}$ (with an outer take-off of $j = j(z, t) + j_i$); the difference is that the non-stationary inner take-off changes with height and is at a maximum at the head of the column: $j_h = j_{\max} = j_{\Omega}(0, t) + j_{\text{s}}$; in the still $j = j_{\text{s}}$ (Fig. 17).

Considering (38c) and (1a) for small concentrations of the component, when $y_e = x/\alpha$ instead of (37), one may write

$$\frac{\partial x}{\partial \overline{z}} + \omega h\frac{\partial x}{\partial t} = \varepsilon x - \left[\omega h\int_{\overline{s}}^N \frac{\partial x}{\partial t}\,d\overline{z} + e_{\text{s}}\frac{\partial x_{\text{s}}}{\partial t} + p\,x_{\text{s}}\right] = \varepsilon x - \frac{i(z,t) + i_0}{L} = \frac{\vartheta}{\lambda}, \tag{37a}$$

where $\overline{z} = z/h$ is the height of the column, measured in transfer units $(\overline{H} = N)$ $\varepsilon = (\alpha\lambda - 1)/\alpha\lambda \approx \varepsilon_0 + p$ is the effective enrichment coefficient, $\varepsilon_0 = (\alpha - 1)/\alpha$, $p = P/L$, P is the amount of product taken off and $\lambda = 1/(1 - P/L) \approx 1 + p$; $e_{\text{s}} = E_{\text{s}}/L$, $\overline{p} = p/\varepsilon_0$.

Differentiating (37a) with respect to $\overline{z}$, we obtain an equation of the second order for $x(z, t)$ [5]

$$\frac{\partial^2 x}{\partial \overline{z}^2} + \omega h\frac{\partial^2 x}{\partial \overline{z}\partial t} - \omega h\frac{\partial x}{\partial t} - \varepsilon\frac{\partial x}{\partial \overline{z}} = 0 \tag{39}$$

and for isotopes with small separation coefficients ($\varepsilon_0 \ll 1$) the second terms in (37), (37a) and (39) may be neglected.

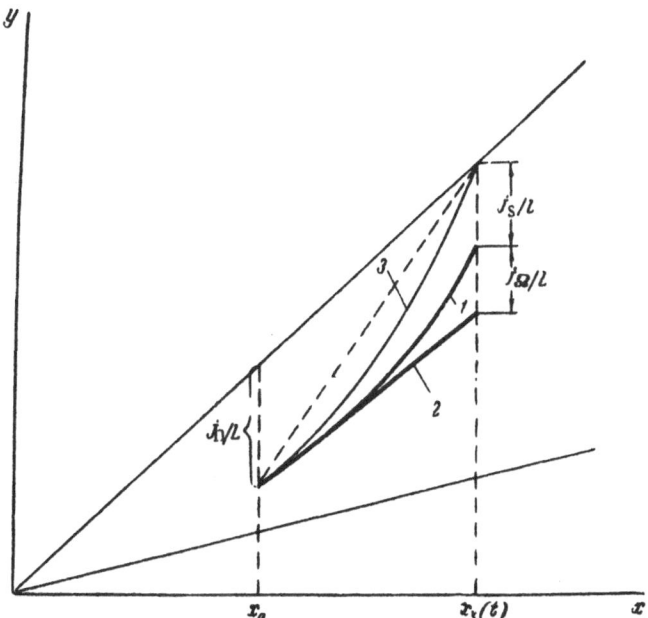

Fig. 17. The x − y diagram of a nonstationary process: j_h and j_s are the nonstationary take-offs in the head and still of the column; 1) operating line of the column with E_s as the holding capacity of the still and that of the packing; 2) schematization of the operating line calculated on the principle of equivalent inner and outer take-offs; 3) operating line of the column, available capacity in the still; dotted line − operating line of a single column, operating with take-off in a stationary state.

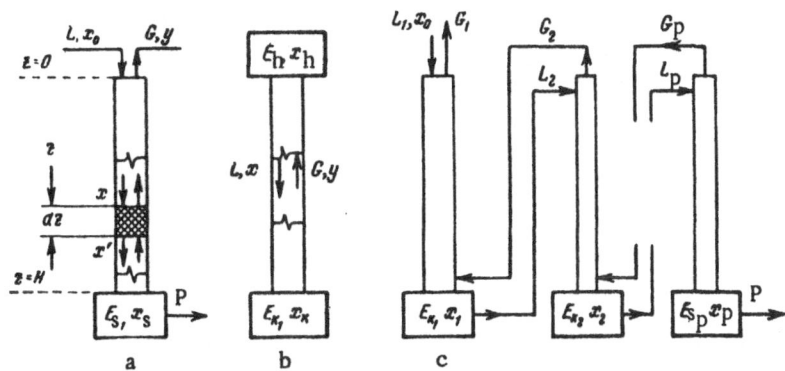

Fig. 18. Some examples of column connections, whose kinetics are being investigated: a) column, fed with a liquid of constant composition, with holding capacity in the still; b) column with two holding capacities: in the still (E_s) and in the head (E_h), without outer feeding; c) cascade of concentrating columns.

2. In the case of a column, fed with a liquid of constant concentration x_0 (Fig. 18a), using an operator method, we find the following form for the solution of Eq. (39): *

---

* A star denotes the representation (transformation) of the value by the Laplace-Heaviside method $x_s^* = p \int_0^\infty x_s(t)\, e^{-pt} dt$.

$$q_s^* - 1 = \frac{x_s^* - x_0}{x_0} = \frac{e^{\varepsilon \Delta r N} - 1}{(\overline{e_s} p - r_2)(e^{\varepsilon \Delta r N} - 1) + \Delta r} = \frac{F_1(p)}{F_2(p)}, \tag{40}$$

where r_1 and r_2 are the roots of the characteristic equation $r^2 - r - \tau_\Omega p = 0$, $\Delta r = r_1 - r_2$, and $\tau_\Omega = \omega h / \epsilon^2$ is a characteristic time for the holding capacity of the packing: $\tau_\Omega = \dfrac{\Omega}{L\epsilon_0 \ln q_0}$.

Analysis of solution (40) shows that at the starting moment ($t \ll \omega h$ $p \to \infty$) accumulation occurs only in the still; if $\overline{e}_s = e_s / \epsilon = E_s / L\epsilon$, then

$$q_s - 1 = \frac{t}{\overline{e}_s + \varepsilon \tau_\Omega} ; \quad x_s - x_0 = \frac{j_0 t}{E_s + \Omega/N} , \tag{40a}$$

however, the duration of this period (of the order ωh — the time the liquid remains on one theoretical plate) is so short that it is practically overlapped by the stabilization period of the hydrodynamic state. The concentration then starts to grow in all plates: $\overline{q}^* = 1/(\overline{e}_s p + \sqrt{\tau_\Omega p})$ and

$$\overline{q}_s = \frac{x_s - x_0}{x_0} = \frac{t}{\overline{e}_s} - \frac{4}{3} \frac{t^{3/2}}{\sqrt{\pi}} \frac{\sqrt{\tau_\Omega}}{\overline{e}_s^2} + \frac{\tau_\Omega}{\overline{e}_s^3} t^2 \ldots \approx \frac{t}{\overline{e}_s + 0.8\tau_\Omega \sqrt{t/\tau_\Omega}} . \tag{40b}$$

Finally, sufficient time after the start of the process, when $t > t_1 = \tau_\Omega \left(\dfrac{\ln q_0}{3}\right)^2$, the solution is reduced to the exponent, whose index p_0 is the least root of the denominator (40). By solving approximately the transcendental equation $F_2(p) = 0$, we find the relaxation time

$$t_0 = -\frac{1}{p_0} = (\overline{e}_s + \tau_\Omega)(q^0 - 1) + 0(\tau_\Omega \ln q^0) \approx \frac{M_p}{j_0} , \tag{40c}$$

where $e_s = \dfrac{e}{\epsilon_0} = \dfrac{E_s}{L\epsilon_0}$ is the characteristic time of the still, $0(\tau_\Omega) \approx 0$, $\epsilon_0 = (\alpha - 1)/\alpha$, M_e is the equilibrium accumulation determined by the relation (38a), and $j_0 = Lx_0\epsilon_0$ is the maximum capacity of the column.

Correspondingly, the general solution to the equation may be written as

$$\overline{q}_s = q_s - 1 = (q_0 - 1)[1 - \delta\psi(t) - (1 - \delta)e^{-t/t_0}], \tag{41}$$

where the main term in the right section is the exponent, $\psi(t)$ is the correction function, which takes into account the starting processes; $|\psi(t)| \leqslant 1$; $\psi(0) = 1; \dfrac{\partial \psi}{\partial t} < 0; \psi(t > t_1) = 0)$, and δ is the amplitude of the correction terms. The latter is not great, and for effective columns ($q_0 > 10$) is less than

$$\delta_0 = \frac{1}{q_0(1+\mu)^2} , \tag{41a}$$

where $\mu = \overline{e}_s / \tau_\Omega = E_s \ln q_0 / \Omega$; when $\mu \sim 1$ $\delta \sim 10^{-2}$. Thus, the amplitude of the correction terms δ is very small (an exact calculation gives $\delta = 0.1$ even with a low-efficiency column, when $q_0 = 3$ and $\mu = 0$) and when $t > t_1$ it may be neglected.[*] As a result, we arrive at a simple kinetic equation when $t > t_1$, which is correct for a still and for any point of a column

$$x - x_0 = (x_p - x_0)(1 - e^{-t/t_0}) \tag{42}$$

or

$$q - 1 = (q_e - 1)(1 - e^{-t/t_0}), \tag{42a}$$

---

[*] More exactly, in the case of efficient columns the deviation δ is very small and may be neglected; δ is considerable in low-efficiency columns (when $q_0 \to 1$, $\mu \to 0$, $\delta \approx 0.2$) but the disappearance time of the correction terms $t_1 \sim (\ln q)^2$ is very small. Thus, in both cases relation (42) is applicable.

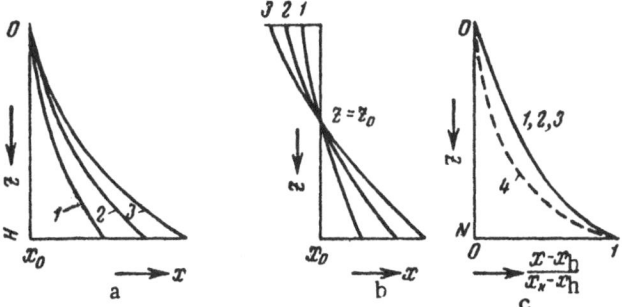

Fig. 19. Concentration cross sections: a) and b) are the columns shown in Fig. 18 a and b; the curves 1, 2 and 3 are when $t_3 > t_2 > t_1$; c) the same in generalized coordinates (spatial similarity); curve 4 is the deviation from similarity when $t < t_1 = \tau 1 (\ln q_0 /3)_2$.

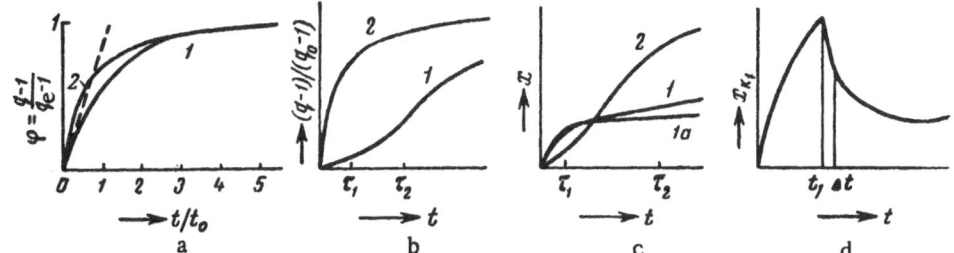

Fig. 20. Kinetic curves: a) curve 1 — columns, shown in Fig. 18 a and b, in generalized coordinates (similarity in time), curve 2 — the same, taking into account the starting processes; dotted line — the initial linear section; b) cascade without flow reduction; 1 — first column; 2 — second column; c) cascade with flow reduction; 1 and 1a — first column with n = 2 and 3; 2 — second column; d) first cascade column; second connected with a delay of t_1.

where $q_e = e^{\epsilon \bar{z}}$; for the still $q_e = q_0 = e^{\epsilon_0 Np} = \alpha^{NT}$. The deviation of (42) from an exact solution of (41) is δe^{-t}, i.e., not only small but also decreases with time. However, the solution is exact for columns with a small holding capacity of packing ($\Omega = 0$) (42) as $\mu = \infty$ and $\delta = 0$.

Equation (42) gives the relation of the concentration of the component being extracted in the column to height (Fig. 19a and b) and to time (Fig. 20); it is applicable only for slow processes as in the case of rapid fluctuations in the load or take-off (with a period $T \ll t_1$), the quasistationary state with exponential kinetics (42) does not have sufficient time to become established.

Passing to a generalized concentration (i.e., introducing instead of $\underline{x}$ the degree of approach to equilibrium φ) we will obtain, instead of (42), a universal kinetic equation, correct for any point of the column (Fig. 20a)

$$\varphi = \frac{x - x_0}{x_e - x_0} = 1 - e^{-t/l_0}, \tag{43}$$

and under these conditions the relaxation time

$$t_0 = \frac{M_e}{j_0} = \frac{\text{equilibrium accumulation of the component in the column}}{\text{maximum capacity of the column}} \tag{43a}$$

The physical meaning of relaxation time is quite simple: it is the time that would be required for attaining equilibrium (i.e., for the accumulation of an equilibrium amount of the component in the still and in the column) if the transport of the isotope were constant and equal to the initial value $j = j_0 = Lx_0\epsilon_0$. Actually, the transport decreases with time, $j = j_0 e^{-t/t_0}$, and to achieve the given degree of approach to equilibrium φ the time required is

$$t = t_0 \ln \frac{1}{1 - \varphi} \tag{44}$$

93

TABLE 2

Conditions of column operation	Equilibrium	
	degree of separation	accumulation
Column with holding capacity in the still, external feeding at constant concentration	$q_e = q_0 = e^{\epsilon_0 N}$	$\dfrac{M_e}{x_0} = E_s\,(q_0 - 1) + \Omega\left(\dfrac{q_0 - 1}{\ln q_0} - 1\right)^{*}$
The same, with external take-off (Fig. 18a)	$q_e = \dfrac{q_0^{1+P}\,(1+P)}{1 + Pq^{1+P}}$	
Column with two reservoirs (in the head and in the still) (Fig. 18b)	$q_e = q_0/\beta^{**}$	$\dfrac{M_e}{x_0} = E_s\left(\dfrac{q_0}{\beta} - 1\right) + \Omega\,\dfrac{\dfrac{q_0}{\beta} - 1 - \ln\dfrac{q_0}{\beta}}{\ln q_0}$

---

*In calculating by the method of transfer units $q_0 = e^{\epsilon_0 N}$; in the case of cascades of single-acting elements and $\alpha < 1$ $q^{*} = \alpha^{N}$; $\epsilon_0 = (\alpha - 1)/\alpha$.

**Here β is the impoverishment of the head reservoir, $\beta = \dfrac{x_0}{x_h} = \dfrac{E_h + E_s q_0 + \Omega(q^{*}-1)/\ln q^{*}}{E_h + E_s + \Omega}$ $\qquad \bar{p} = p/\epsilon_0 =$ $= P/L\epsilon_0$; and E_h is the capacity of the head reservoir.

Correspondingly, a 98% approach to equilibrium requires 4 t_0 and 99% — 5 t_0 (Fig. 20a; the dotted line shows the approach of A. I. Brodskii [1] according to whom $t = t_0$). It may be readily seen that Brodskii's approximation corresponds to the initial linear section of the kinetics, when $\varphi = t/t_0$ and $x_s - x_0 = j_0 t/E_s$. The equation, however, for calculating the time for attaining the given concentration, obtained by S. I. Babkov and N. M. Zhavoronkov by an approximation method [6] may be reduced to (44). For laboratory columns with low efficiency, when $q^0 - 1$, < 1, $t \approx 5\ NE_s/L$.

3. Investigation of a series of column connections showed that the kinetic Eqs. (42)–(44) and determination of relaxation time (43a) have a general character. They are correct for various operating conditions and column connections. The values used in these equations for the parameters of different cases of column operation are given in Table 2.

The relations (42)–(43) are applicable in the case of a column with two reservoirs as such a column has a "zero point" (at the height $\bar{z}_0 = \dfrac{1}{\epsilon_0} \ln \beta$) in which the concentration does not change with time and is equal to the initial one (Fig. 19b). The lower part of the apparatus works, therefore, as a column which is fed at a constant concentration. Correspondingly, the accumulation M_e is calculated for the lower part of the column (from height $N - \bar{z}_0$), as $\bar{\Omega} \int_{z_0}^{H} (x - x_0)\,dz + E_s\,(x_s - x_0)$.

In the case of a cascade, the universal kinetic Eqs. (42)–(43) are also correct for the last column, with M_e — the accumulation in the cascade, j_0 — the maximum capacity of the cascade $j_0 = j_1^0/R$, where j_1^0 is the maximum capacity of the first column, $j_1^0 = L_1 x_0 \epsilon$, and R is the cascade resistance. However, for the first column (in the case of a cascade of two columns) $\varphi_1 \approx 1 - (1 - \theta_1)\,e^{-t/\tau_1} - \theta_1 e^{-t/t_0}$, where θ_1 is the relative take-off in a stationary rated condition; during linear kinetics of the last column ($t \ll t_0$) the preceding columns operate in a quasirelaxation state, quite close to the rated stationary condition with take-off (Fig. 20c); τ_1 is also determined by Eq. (43a), with M_{e1} — the accumulation in a quasistationary state, $M_{e1} = (q_1^0 - 1)(1-\theta_1)x_0 E_{s1}$.

If, as usually occurs, the rated state of the cascade is far from equilibrium without take-off ($Q/Q_0 \ll 1$), then the period of accumulation may be determined from the accumulation and capacity in the rated state,

$$T = \left(\sum_{i=1}^{n} M_i\right)\bigg/ (j_1^0/R),$$ where M_i is the accumulation in the i-th column. The holding capacities of the

packing or plates of the column, required for calculating accumulation, may be determined by the pulse method based on nonstationary processes (the time $\bar{\tau}$ is found during which a pulse, for example from some breakdown in the liquid's movement, passes from the head to the still of the column; $\Omega = L\bar{\tau}$; if the emergent pulse is diffuse,

then $\bar{\tau} = \int_0^\infty \tau i(\tau)\, d\tau \Big/ \int_0^\infty i(\tau)\, d\tau,$ where $i(\tau)$ is the intensity of the emergent curve; $\tau = 0$ at the moment

when the pulse is introduced at the head of the column).

The relations (42)-(43), somewhat modified, are also correct for nonsimultaneous connection of the columns or of take-off; different cases, shown in Fig. 18, are described by the equation $q - q_i = (q_e - q_i)(1 - e^{-t/t_0})$, where q_i and q_e are the initial and the equilibrium degrees of separation.

In the case of high concentrations of the required component, when $x > 0.9$, Eqs. (37)-(39) may be solved for the volatile isotope, which again gives kinetic relations (42)-(43), with $\varphi = 1 - e^{-t/t_0}$, where $\varphi = \overline{\overline{x}}_e = \overline{\overline{x}}_0 e^{-\epsilon_0 \overline{z}})$. In this case, the time of relaxation is determined not by the accumulation of the low-volatility component, but by the removal of the volatile one: $\overline{\overline{t}}_0 = \overline{M}_e / \overline{J}_0 = M_e / j_0 (1 - x_0) = t_0 / (1 - x_0)$, as the maximum capacity with respect to the volatile component $\overline{j}_0 = L\epsilon(1 - x_0)$ is much less than that for the nonvolatile one [if one examines a column which concentrates the volatile isotope, then relations (42)-(43) are completely correct for it].

4. Integration of the general equation of mass transfer (37a) with the assumption that the nonstationary internal take-off j is only a function of time t (but not of coordinate z),* which is exact for a column with a small holding capacity of the packing and is a good approximation when $\Omega \neq 0$, makes it possible to establish two important principles [5]:

(a) the internal nonstationary and the external stationary (cascade) take-offs are equivalent — they decrease concentration head and enrichment in the same manner (Fig. 17); consequently, Eq. (17a) is correct for a nonstationary process

$$q - 1 = (q^0 - 1)(1 - j/j_0); \quad \varphi = \frac{q-1}{q^0-1} = 1 - j/j_0 \tag{45}$$

(in this case, j means the full nonstationary take-off j_h which somewhat decreases the concentration head; however, as Fig. 17 shows, the approximation is more exact than Formula (17a) for a single column, see comment on p. 82).

(b) The stationary and nonstationary concentration cross sections in the column and the kinetic curves for any point of the column are similar (in the former case — spatial similarity, in the latter — time similarity), Fig. 19c and 20a:

$$\frac{x - x_0}{x_s - x_0} = \frac{x_e - x_0}{x_{se} - x_0} = \frac{e^{\epsilon_0 z} - 1}{e^{\epsilon_0 N} - 1} = f(z); \tag{46}$$

$$\frac{x - x_0}{x_e - x_0} = \frac{q(z,t) - 1}{q_e(z) - 1} = \varphi(t). \tag{47}$$

Thus, in the approximation examined [Eq. (47) is equivalent to Eq. (42)] the degree of approach to equilibrium φ is the same for all the plates of the column.

Using these two principles one may arrive at kinetic Eq. (43a) by an elementary method. From (47) and (38) it follows that $M = M_e \varphi$, so that $j = dM/dt = M_e d\varphi/dt$. As from (45) $j = j_0(1 - \varphi)$, then

$$M_e \frac{d\varphi}{dt} = j_0(1 - \varphi) \tag{48}$$

$$\varphi = 1 - e^{-tj_0/M_e} = 1 - e^{-t/t_0}, \tag{49}$$

which coincides with relation (43). A cascade of columns may also be examined in a similar way.

By taking the value $q_0 - 1$ out of the braces in Eq. (38a), one can show that the simplification introduced by Eq. (47) indicates that the holding capacity of the packing distributed throughout the height is substituted by an equivalent holding capacity

---

* Which is equivalent to using the quasistationary state principle.

$$\Omega_{eq.} = \Omega \Big/ \Big(\frac{1}{\ln q_0} - \frac{1}{q_0 - 1} \Big),$$

in the still of the column. Expanding (42a) into a series, at $t \ll t_0$, we find that this substitution leads to the relation $x_s - x_0 = \dfrac{j_0 \cdot t}{E_s + \omega_{eq}}$, according to which the equivalent holding capacity of the packing participates all the time in the accumulation; actually, however, the holding capacity of the packing is involved in the accumulation gradually, increasing by Eqs. (40a) and (40b) and time $\underline{t}$, from one plate to Ω_{eq} .

The principle of similarity is useful for solving a series of problems: in particular, for determining the number of plates in a column without waiting for equilibrium. For this it suffices to measure the concentration at three points of the column — for example, at the head, the still, and in one intermediate point $\underline{z}$. As $q_e(z) = e^{\epsilon z} = q_0^{zH}$, the similarity [Eq. (46)] gives $(q_0^{zH} - 1)/(q_0 - 1) = (x - x_h)/(x_s - x_h)$, from which q_0 and $N = (\ln q_0)/\epsilon_0$ may be found.

In the region of intermediate concentrations $(0.15 < x < 0.85)$ Eq. (37) must be made linear, for example, by Cohen [14]. One can show that in this case as well, with certain approximations, the kinetics will be described by Eq. (42), if the absolute concentrations $\underline{x}$ are replaced by the relative ones $X = x/(1-x)$, i.e., $X - X_0 \approx (X_e - X_0)(1 - e^{-t/t_0})$. Thus, the establishment time is determined by both factors: removal of the volatile component and accumulation of the heavy one.

The principle of similarity remains correct for high concentrations of isotopes; for this one should replace the absolute concentrations $\underline{x}$ in Eq. (43) by the relative ones X (we indicated above that in relative concentrations the cross section through the height of the column does not differ from the case when $x \ll 1$; cf. Figs. 6 and 11).

LITERATURE CITED

[1] A. I. Brodskii, Chemistry of Isotopes (Acad. Sci. USSR Press, 1952).

[2] K. Cohen, Sci. and Eng. Nucl. Power (Ed. C. Goodman) (New York, 1949); Theory of Isotope Separation (New York, 1951).

[3] N. M. Zhavoronkov and Ia. D. Zel'venskii, Collection: Processes and Apparatuses of Chemical Technology (State Chem. Press, 1953).

[4] A. M. Rozen, Doklady Akad. Sci. USSR 108, 122 (1956).

[5] A. M. Rozen, Doklady Akad. Sci. USSR 107, 295 (1956).

[6] N. M. Zhavoronkov and S. I. Babkov, Doklady Akad. Sci. USSR 106, 877 (1956).

[7] A. Kasatkin, Basic Processes and Apparatuses of Chemical Technology (Moscow, State Chem. Press, 1955).

[8] M. Fenske, Ind. Eng. Chem. 24, 482 (1932).

[9] A. Colburn and T. Chilton, Ind. Eng. Chem. 26, 1183 (1934).

[10] N. Gel'perin, Distillation and Fractionation (Moscow, State Sci. Tech. Press, 1947).

[11] A. Stoppel, Ind. Eng. Chem. 38, 1271 (1946).

[12] E. Becker, Angewandte Chemie 68, 1 (1956).

[13] N. Kuz'minykh, and S. Kalinina, J. Appl. Chem. (1951).

[14] K. Cohen, J. Chem. Phys. 8, 588 (1940).

SEPARATION OF ISOTOPES OF LIGHT ELEMENTS BY DIFFUSION IN VAPORS

I. G. Gverdtsiteli and V. K. Tskhakaia

Diffusion is widely used for the separation of isotope and gas mixtures. Some diffusion processes, for example gaseous diffusion and thermal diffusion, have been studied quite well and find a wide application not only in laboratory practice, but also in the industrial preparation of U^{235} (gaseous diffusion).

Among the diffusion methods of separation, there is also the process of diffusion in a vapor jet (so-called counter-current diffusion). The first papers in this field appeared in 1922-1923, when G. Hertz [1] used a steam jet as a diffusion medium and completely separated a neon-helium mixture. In 1934-1939, a whole series of papers [2-6] appeared on further investigation of the separation process and developments in the construction of the separating elements. In 1934, in an experimental model, Hertz was able to combine the separating element and the circulating pump into one apparatus (separating pump) and thus it became possible to realize a cascade process without auxiliary circulating pumps. The isotopes of hydrogen, neon, argon, carbon, etc. were separated on cascade apparatuses consisting of 6 to 48 pumps. The working liquid used was mercury.

The main disadvantage of separating pumps was the very low capacity, which made it possible to obtain only indicator amounts of highly enriched isotopes. The construction of the pump could not be used as the basis for building high-capacity separating apparatuses. In our investigations particular attention was paid to the construction of models, which could serve as a basis for developing high-capacity separating apparatuses.

From 1946 to 1952 the authors of the present article worked under the direction of Hertz on an investigation of the separation process and the construction of high-efficiency apparatuses with a high capacity. The experimental results presented were mainly obtained in 1956.

Essential Plan of Separating Pump

The essential plan of the separating pump is given in Fig. 1. The separating pump consists of four main sections: the evaporator 1, the diaphragm 2, and two condensing surfaces 3 and 4. Vapor from the evaporator 1 passes into the space I, which is surrounded by a porous diaphragm. One part of the vapor passes through the openings of the diaphragm and is deposited on the condensing surface 3. The transverse flow of the vapor forms a diffusion space in which the separating process is realized. The other part of the vapor flows along the diaphragm and precipitates on condenser 4. The liquid from the two condensers drains through a liquid seal back into the evaporator. The separable isotope mixture passes into region II (between the diaphragm and the condenser).

Considering that near the condensing surface 3 the partial vapor pressure is extremely small (adjacent to it the space will be filled with gas), and that in region I there will be mainly vapor, gas diffusion against the vapor flow arises. It is obvious that the part of the gas which diffuses through against the current will be enriched in the light component (in the general case, the component with the greatest diffusion coefficient). The gas diffusing through into region I will be caught by the longitudinal vapor flow and be carried to condenser 4, where it will be freed from vapor.

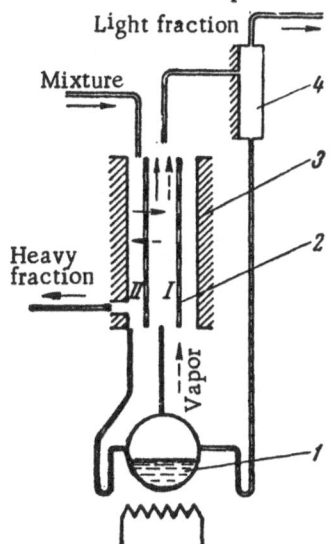

Fig. 1. Essential plan of separating pump.

97

The isotopic mixture entering the separating pump will be continuously impoverished in the light isotope as it moves along the condenser and will emerge from the pump as the heavy fraction. Thus, the separating pump divides the stream of mixture into two parts — a light fraction stream and a heavy fraction stream. Since the diaphragm offers a certain hydrodynamic resistance to the flow of vapor across it, the over-all pressure in region I will be slightly greater than in region II. This effect is used to circulate the streams between pumps in the cascade scheme.

Main Parameters of the Process

A diffusion process in a vapor stream is a problem of convection diffusion for three components. For simplicity we will assume that the components of the gas participate in the diffusion process independently of each other (the case of low concentrations). The conclusions based on this assumption are the first approximation of the more rigid theory and give a satisfactory qualitative picture of the process.

The convection diffusion equation for the one-dimensional case (diffusion along the x axis) will have the form:

$$\tau = - NS \left[D \frac{d\gamma}{dx} + U\gamma \right],$$ (1)

where τ — the gas flow; D — the diffusion coefficient for the gas in the vapor; γ — the concentration of the gas in the vapor; U — the speed of the mixture; S — the area of an opening; N — the number of openings in the diaphragm.

The solution of this equation will be:

$$\gamma(x) = \left(\gamma_0 + \frac{\tau_0}{SU} \right) e^{-\frac{Ux}{D}} - \frac{\tau_0}{SU},$$ (2)

where $\tau_0 = \tau / N$ is the gas flow through one opening.

From (1) and (2) it follows that

$$\tau = NS \frac{D\gamma_0}{l} \frac{\ln q}{q-1} \left(1 - q \frac{\gamma_l}{\gamma_0} \right),$$ (3)

where γ_0 and γ_1 are the concentrations of gas in vapor at the beginning and end of the diffusion path l :

$$\ln q = \frac{Ul}{D} = \text{Pe}_D \text{ is the Peclet diffusion number.}$$

From (3) it follows that the flow of gas diffusing through (circulation) is an exponential function of Pe_D. From (3) one can obtain an expression for the separation coefficient

$$\alpha = \frac{q_2 - 1}{q_1 - 1}.$$ (4)

For cases when $\alpha \approx 1 + \epsilon$, where ϵ differs from unity, we obtain

$$\varepsilon = 1{,}386 \frac{q \ln q}{q-1} \frac{\delta D}{D},$$ (5)

where $\delta D/D$ is the relative difference in the diffusion coefficients of the isotopes in the vapor; 1.386 is the Rayleigh coefficient for the case when the emergent stream is divided into two equal parts.

For large values of q, $\epsilon \sim \ln q$, i.e., the enrichment factor is proportional to the speed of the vapor.

As has already been indicated, the diaphragm is the main element of the separating pump. In our experiments, the diaphragm was a cylinder with from 20 to 20,000 openings per cm² of surface and opening diameters of from 0.5 to $5 \cdot 10^{-3}$ mm. The diameter of the openings had to be considerably larger than the mean free path of the molecules. To achieve efficient separation and steady conditions for the pump, it was important to match the hydrodynamic and diffusion resistances of the diaphragm reasonably. The correct choice of working liquid was also very important. The vapor of the liquid had to be neutral to the gases being separated and to the materials used in the construction of the apparatus. The residual pressure of the vapor at the condensation temperature could not be more than 10% of the total pressure. It was desirable to use a liquid in which the solubility of the gas separated was minimal because transference by solubility was parasitic and led to a reduction in the efficiency of the separation stage. There are particular requirements as regards the molecular weight. From (5) it follows that $\epsilon \sim \delta D/D$. The value of $\delta D/D$ tends to a maximum equal to $\Delta m/2m$ at $M \rightarrow \infty$ (M is the molecular weight of the vapor). Practically it was necessary that the molecular weight of the vapor was 3-4 times greater than the molecular weight of the gas. For solving different problems we used heptane, xylene, ethyl alcohol, and mercury.

Results of Separating Isotopic Mixtures

The separating pumps were made up into an experimental block to investigate their main characteristics. The number of pumps in the block was determined by the accuracy requirements of the concentration measurements. The concentration measurements were performed with an MS-type mass spectrometer. The values of the interstage flows were measured with special flow meters.

1. Small Laboratory Model

With the exception of the diaphragm, the separating pumps were made entirely from glass. The diaphragm was a steel tube with a diameter of 15 mm and a wall thickness of 0.3 mm. The diaphragm surface was pierced with 500 openings, 0.4 mm in diameter. The total area of all the openings was 0.6 cm^2. The height of the working part of the diaphragm was 40 mm. The diaphragm was attached to glass tubes at the two ends. The condensing surface was 3 mm from the diaphragm. The internal condensing surface was cooled with running water. The second condenser was air-cooled.

Figure 2 illustrates the dependence of the enrichment factor (curve 1) and the transfer (curve 3), obtained in experiments on the separation of neon isotopes, on the value of the dimensionless Pecletparameter (Ul/D). The experiment was performed at 10 mm Hg pressure. As an illustration, the electrical power of the evaporator is plotted along the abscissa. The dotted curve 2 corresponds to the theoretical limiting value of the circulation at $\gamma_0 \rightarrow 0$. The theoretical curve corresponds to practical data for $l_{eff} = {}^5/_3\, l_{geom}$. This empirical rule is very important as it shows that the diffusion process is not only limited to the openings of the diaphragm, but also occurs in the layers adjacent to the diaphragm. This idea is confirmed by many experimental results. In addition, we should note that the ratio of the effective length of the diffusion path to the geometrical thickness of the diaphragm depends on the form of the openings. Thus, for example, for slot diaphragms (slots 0.3 mm wide and 40 mm long) $l_{eff}/l_{geom} \approx 2$.

It is evident that in a more strict treatment l_{eff} would also depend on the speed of the vapor through the diaphragm. However, this dependence would be weak.

The presence of a maximum at $Pe_D = 1$ is due to the fact that at $Pe_D < 1$ the concentration of gas in the vapor in region I, γ_e reaches high values because of the low value of the longitudinal component of the vapor.

The maximum separation coefficient in experiments with neon isotopes was 1.2 ($Pe_D \sim 3$). No further increase in the separation coefficient was possible due to the limited power of the evaporator.

The region where $Pe_D > 4$ is not of great practical interest due to the low capacity of the pump.

Most of the experiments were performed at 10 mm Hg pressure. Investigations were carried out in the region 8-50 mm Hg. It was established that in this pressure range the value of the separation coefficient did not depend on the pressure and was determined by Pe_D only.

The separation coefficients and circulations for some isotopic compounds are presented below.

Figure 3 gives the separation coefficients for different isotope mixtures as functions of the relative difference in the diffusion coefficients. Using this curve, one can evaluate with sufficient accuracy the possible separation coefficients for the range $0.005 \le \delta D/D \le 0.05$.

Preliminary results, obtained on blocks of 4-5 separating pumps, showed that it was worthwhile building a large cascade.

Figure 4 is a photograph of a cascade composed of 70 pumps. The separating pumps are arranged in two rows. All the sections of the cascade were made from molybdenum glass. The total length of the cascade is 5 m. Highly enriched isotopes of argon, krypton, oxygen, and carbon were obtained on the cascade.

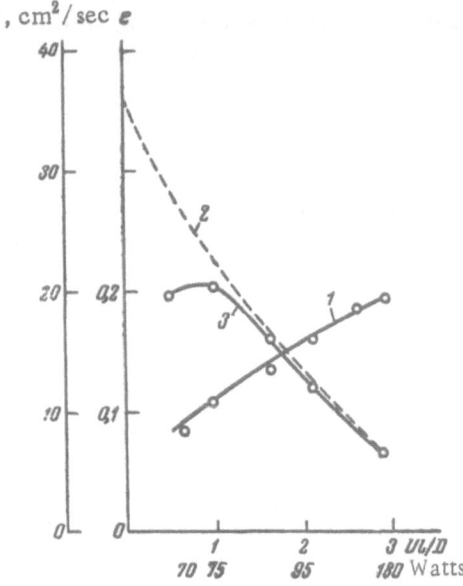

Fig. 2. Dependence of enrichment factor e and transfer τ on Ul/D.

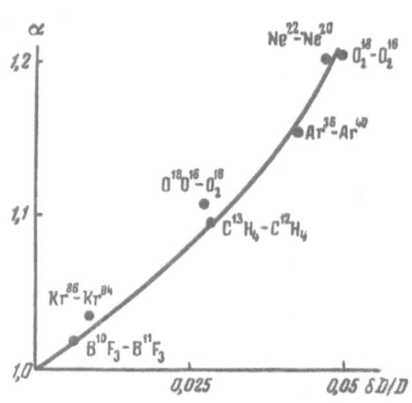

Fig. 3. Dependence of separation coefficient on relative difference in diffusion coefficients δD/D.

Fig. 4. Cascade of 70 pumps.

Fig. 5 shows the course of the change in concentration of $C^{13}H_4$ along the cascade with various take-off rates. The starting material used was methane containing 2.7% C^{13}. The work was performed at a pressure of 10 mm Hg. The fall in pressure along the cascade was about 1 mm Hg. About 80 n-cm^3 of gas was required to fill the cascade.

So as to use the whole separation effect for enrichment, the cascade was operated without a stripping section. To the light end of the cascade was attached a special evaporation-condensation system which gave a circulation interchange with the first pump. The maximum C^{13} concentration at a take-off rate of 30 cm^3/day was 88%. The over-all enrichment on the cascade was 263, i.e., an average of 1.086 per stage. At the maximum take-off rate of 200 n-cm^3/day, the separation coefficient was 26.

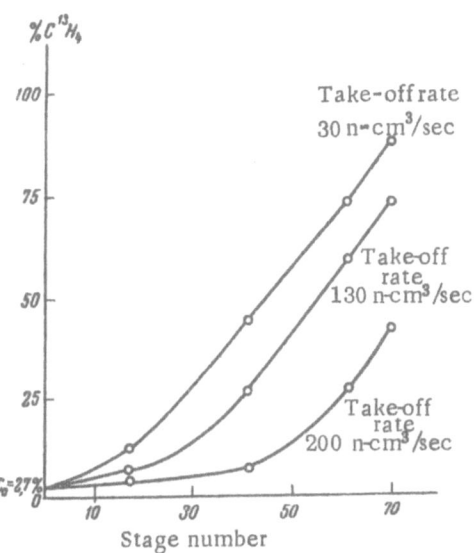

Fig. 5. Course of $C^{13}H_4$ concentration along a cascade with different take-off rates.

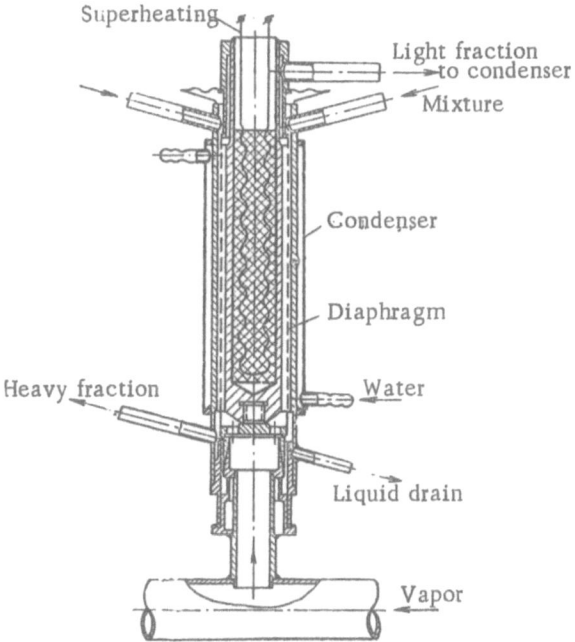

Fig. 6. Plan of copper separating pump.

TABLE

Isotope mixture	Separation coefficient	Circulation n-cm³/sec
$Ne^{22} - Ne^{20}$	1.2	0.26
$Ar^{40} - Ar^{36}$	1.15	0.12
$C^{13}H_4 - C^{12}H_4$	1.095	0.28
$B^{11}F_3 - B^{10}F_3$	1.016	–
$Kr^{86} - Kr^{84}$	1.033	0.05
$O^{16}O^{18} - O_2^{16}$	1.105	0.18
$O_2^{18} - O_2^{16}$	1.21	0.18

2. Separation Stages of High Capacity

A small laboratory cascade can be used most effectively at high concentrations (i.e.,where the capacity required is comparatively small). For the preliminary enrichment of isotopes we constructed a cascade from separating pumps of a considerable capacity. Fig. 6 gives a schematic plan of a separating pump. The pump was readily dismantled. It was made entirely of copper. All the inner surfaces were carefully nickel plated. The diaphragm was made from nickel gauze with 20,000 openings per cm², which was given a preliminary treatment to reduce the opening size to 10^{-2} mm. The total area of the diaphragm was 300 cm². Xylene and ethyl alcohol were used as working liquids. While all the basic elements and internal geometry of the laboratory model was retained, in the new model the diaphragm was specially heated to prevent the vapor of the working liquid condensing. A cascade apparatus consisting of 40 separating pumps was investigated for the separation of argon isotopes (the working liquid used was ethyl alcohol). The circulation equalled 6.6 n-cm³/sec and the maximum separation coefficient with material withdrawal was 1.12 per pump (with a take-off of 0.5 n-liter of argon per day). The operating pressure was 1 atm.

Cascade apparatuses, consisting of diffusion pumps, were very stable in operation; they required only periodic inspection. Separating apparatuses have recently been developed which use secondary separating effects. The pumps are fitted with diaphragms 200 mm long. The total number of openings in the diaphragm is 5000 and the opening diameter is 0.3 mm. The working liquid is mercury. In the separation of neon isotopes on the new pumps, the separation coefficient reached 1.6 and for carbon isotopes (in methane) it reached 1.26.

From the experimental results obtained in an investigation of various types of separating apparatuses employing diffusion in a vapor jet, it was established that cascades of diffusion pumps could be used successfully for the preparation of highly enriched isotopes of a series of elements over a wide mass range.

The specific separation capacity of the pumps investigated above considerably exceeds (by approximately 80-100 times) that of pumps described in the literature.

The main part of the mass spectrometric measurements was performed by K. V. Kuprianov and L. I. Chernova under the direction of K. G. Ordzhonikidze.

LITERATURE CITED

[1] G. Hertz, Z. Physik. 23, 433 (1922).

[2] G. Hertz, Z. Physik. 91, 810 (1934).

[3] H. Barwich, Z. Physik. 100, 166 (1936).

[4] Kopferman and Krüger, Z. Physik. 105, 319 (1937).

[5] Sherr, J. Chem. Phys. 6, 251 (1938).

[6] Hemptinne and Capron, J. Phys. et Radium 10, 4, 171 (1939).

A DIFFUSION COLUMN FOR THE SEPARATION OF ISOTOPES

G. F. Barvikh and R. Ia. Kucherov

Up to the present time several papers [1-3] have been published in the literature on the experimental investigation of diffusion columns (separating columns based on diffusion in a vapor current). In only one of these papers [3] was an attempt made to use a diffusion column for isotope separation. Judging by the results presented, the apparatuses described in these papers can hardly be used for the practical preparation of isotopes.

A series of Soviet authors carried out work leading to the construction of efficient separating devices of this type.

Below we present a description of a laboratory diffusion separating column used for the separation of isotopes of some light elements. A cascade of these columns may be used for the preapration of highly enriched isotope mixtures.

Operating Principle and Design of Apparatus

Let us consider a cylindrical vessel, inside which is placed a porous tube along the axis, and through this, vapor is admitted to the vessel. The vapor condenses on the walls of the vessel and passes out of the apparatus. If the vessel is filled with a gaseous mixture of two isotopes, which we will subsequently refer to as the gas for short, then the flow of vapor drives the gas into the condenser and a concentration gradient arises in the gas-vapor mixture across a section of the apparatus.

The presence of a concentration gradient causes diffusion currents, tending to equalize the concentrations in the mixture. In the end a stationary concentration distribution will be established in the vessel in which the convection current of the gas, caused by entrainment by the vapor moving to the condenser, equals the contrary diffusion flow of the gas. Due to the difference in the diffusion properties of the isotopes, the concentration of the light, more mobile isotope will be slightly higher in the porous tube than in the condenser, i.e., an isotope separation occurs. The separation coefficient in this, so-called elementary process, is small.

However, in the same apparatus it is possible to obtain a considerable change in the isotope concentration. This only requires that simultaneously with the imput of vapor, longitudinal convection currents are created [4], one carrying the gas in the porous tube upwards and the other in the condenser, downwards. These currents disrupt the stationary distribution described. Due to the mixing caused by this, the isotope concentration gradient in the cross section is slightly decreased, but on the other hand there arises a current of the light isotope towards the porous tube and a counter current of the heavy isotope. As a result of this the ascending gas stream will be enriched in the light isotope and the descending stream will be impoverished in it, which makes it possible to obtain a considerable difference in the isotope concentration of the gas at the top and the bottom ends.

In the practical realization of this process in an apparatus (Fig. 1), a tubular porous diaphragm 4 was introduced to separate the ascending and the descending streams. Instead of a porous tube for supplying vapor to the apparatus, a metal tube 3 was used, and this had a series of openings drilled in it at different heights and at equal distances from each other.

The presence of the diaphragm not only prevented undesirable mixing, but made it possible to create externally regulated longitudinal gas streams. For this purpose a small additional amount of vapor was introduced at the bottom of the apparatus into the annular space 5, situated between the diaphragm and the vapor imput

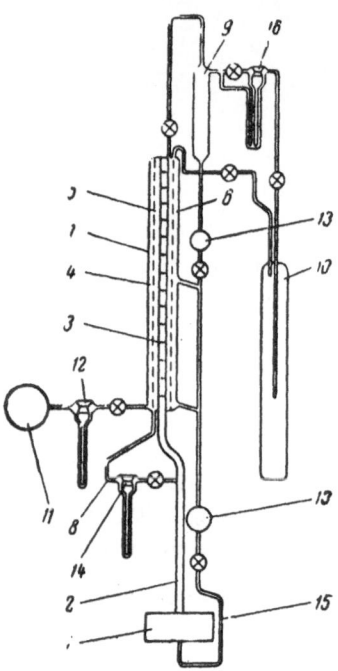

Fig. 1. Plan of experimental apparatus: 1) evaporator; 2) vapor imput; 3) vapor supply tube; 4) tubular porous diaphragm; 5) annular space; 6) space between diaphragm and condenser; 7) internal condenser; 8) tube for additional vapor imput into lower part of column; 9) external condenser; 10) reservoir; 11) bulb for gas withdrawn; 12) capillary flowmeter to measure take-off; 13) equipment to measure liquid flow; 14) and 16) capillary flowmeters; 15) tube for drainage of liquid into evaporator; circles with crosses inside represent valves.

tube, and an approximately equal amount of a gas-vapor mixture was removed from the upper part of this space into an external condenser 9. The hydrodynamic current thus produced in space 5 provided gas transfer upwards along the diaphragm. Gas was led from the external condenser through the capillary flowmeter 16 and the reservoir 10, and introduced into the upper part of the apparatus into space 6 between the diaphragm and the condenser, whence it passed downwards along the diaphragm. In the lower part of the apparatus the gas was returned into space 5 by diffusion through the diaphragm in a stream of fresh vapor introduced into the column. The flow of gas through the external condenser was caused by the hydrodynamic fall in pressure at the diaphragm, produced by the vapor passing through the diaphragm.

The imput of vapor into the apparatus with a tube with a uniform distribution of openings along it, led to some complication in the movement of the gas. In each small portion of the apparatus, sited between successive series of openings (subsequently, such a portion will be referred to as a section), gas currents arise through the diaphragm directed into the lower part of the section from the condenser and in the upper part of it, towards the condenser. As a result there is some modulation of the longitudinal gas stream.

The column was made from stainless steel. The length of the diaphragm was 1 m and the diameter 38 × 40 mm. The diffusion resistance of the diaphragm [5] was approximately 0.5 cm. The vapor supply tube had 24 series of openings and 15 openings 1 mm in diameter in each series. The separation between the vapor supply tube and the diaphragm was 4 mm and that between the diaphragm and the condenser, 2.5 mm. To prevent condensation of vapor on the diaphragm and the vapor supply tube, a small electrical heater was placed inside the latter (not shown in Fig. 1). The top and the bottom ends of the column had small devices (also not shown in the figure) for withdrawing small samples of gas into glass ampules for subsequent mass spectrometric measurements.

The accessory vapors used were those of xylene, nitrobenzene, and some other organic liquids, whose molecular weights exceeded severalfold the molecular weight of the gas separated.

Theory

The separation process in the column is based on the use of a diffusion phenomenon in multicomponent gas mixtures. We will limit ourselves to examining the separation of a mixture of two isotopes using a vapor which plays an accessory role. In addition we will neglect the effects of thermal and pressure diffusion, which have little effect on the operation of the column.

The system of equations describing, in the first approximation, the kinetic theory of diffusion of multicomponent and, in particular, three-component gas mixtures [Eq. (25) in [6]], may be simplified in this case. Neglecting terms in $\epsilon = (D_{13} - D_{23})/D_{13}$, where D_{13} and D_{23} correspond to the diffusion coefficients of the light and heavy isotopes in the vapor, respectively, in comparison with unity, after some rearrangement we obtain the two following relations:

$$\vec{\tau} - \vec{u}\gamma = -nD_{13}\nabla\gamma, \tag{1}$$

$$\vec{\tau}_1 - \vec{\tau}c = -Dn\gamma\left\{\nabla c - \varepsilon c\,(1-c)\,\nabla \ln\gamma\right\}, \tag{2}$$

where

$$D = \frac{D_{12}D_{13}}{D_{13}\gamma + D_{12}\,(1-\gamma)}\,;$$

$\vec{\tau}_1$ – the current density of light isotope molecules; c – the mole fraction of the light isotope in the gas; $\vec{\tau}$ – the current density of the gas molecules; γ – the mole fraction of the gas in the gas-vapor mixture; $\vec{u}$ – the current density of the mixture of gas and vapor molecules; D_{12} – the interdiffusion coefficient of the light and heavy isotopes.

Let us take these equations and the continuity equations to investigate the column. We will limit ourselves to considering the stationary plane problem, and following [7], will assume that the diffusion process in the laminar stream of gas-vapor mixture on the two sides of the diaphragm may be allowed for by adding to the diffusion resistance of the diaphragm l the value $(13/35)(a + b)$, where a is the width of the gap between the condenser and the diaphragm, and b is the width of the gap between the diaphragm and the vapor supply tube. Also considering the fact that the difference between the concentrations of the light isotope on the two sides of the diaphragm is small in comparison with the concentrations themselves, we may obtain with the help of (2) the following equation describing the transfer of the light isotope in the column:

$$I_1 = \frac{G\alpha}{\delta}\,c\,(1-c) - \left(\frac{1-\delta}{\delta\tau_x}\,G^2 + R_D\right)\frac{dc}{dz} + Ic, \tag{3}$$

where

$$\alpha = \int_0^s \varepsilon\,\frac{\partial \ln\gamma}{\partial x}\,e^{-\tau_x\int_0^x \frac{dx}{D\gamma}}\,dx, \tag{4}$$

$$\delta = e^{-\tau_x\int_0^s \frac{dx}{D\gamma}} \tag{5}$$

$$s = l + \frac{13}{35}\,(a+b), \tag{6}$$

$$R_D = aD_a\gamma_a + bD_b\gamma_b. \tag{7}$$

The origin of the coordinates is on the side of the diaphragm closest to the condenser. The z axis is parallel to the diaphragm and the x axis perpendicular to it and directed into the diaphragm.

G is the longitudinal gas current along one side of the diaphragm, relative to the diaphragm perimeter; I and I_1 are, respectively, the take-off of gas and light isotope from the apparatus, also relative to the perimeter of the diaphragm; τ_x is the current density of the gas through the diaphragm.

The index a or b in (7) indicates that the value in question refers to the space between the diaphragm and the condenser or the space between the diaphragm and the vapor supply tube.

With the presence of mixing, caused by parasitic convection currents, extra terms appear on the right side of (3) with the form $R_p\,(dc/dz)$.

We should emphasize that in the derivation of Eq. (3) no assumptions were made as regards the dependence of τ and γ on z. Therefore, the equation can be used to investigate apparatuses in which the relation $\tau_x \equiv 0$, which is usually postulated in deriving equations for separating columns, is not fulfilled.

The coefficients in Eq. (3) may be calculated by solving the problem of the motion of the gas in the column. This was done on the assumption that the current density of the molecules of the gas-vapor mixture through the diaphragm u_x did not depend on z. (Since the fall in pressure across the diaphragm is large in comparison with the longitudinal falls in pressure, and γ_a is determined by the condensation conditions and depends very little on z, then in a uniform diaphragm insignificant changes in u_x with changes in z may arise only as a result of the dependence of γ_b on z.)

The solution of this problem showed that if we do not consider the first part of the apparatus, usually not more than 10-15% of the length of the apparatus, the gas stream along the diaphragm G and also γ_b and τ_x depend very little on the original conditions, i.e., on the gas currents in the first and the last sections. Their values are determined completely by the value of the hydrodynamic current in the space between the diaphragm and the condenser Q(z) and the Peclet diffusion number, characterizing the flow of the gas-vapor mixture in the diaphragm:

$$\text{Pe}_D = \frac{u_x s}{D}.$$

(8)

Since Q(z) is the periodic function of $\underline{z}$ with a period equal to the length of the section, L, then G, γ_b and τ_x and, consequently, the coefficients in Eq. (3) also are periodic functions of $\underline{z}$ with the same period.

One can show that the solution of Eq. (3) with periodic coefficients, satisfying the condition $c = c_0$ at $z = 0$ with sufficient accuracy for practical purposes, may be replaced by the solution of the equation of a rectangular cascade, satisfying the same limiting condition

$$\frac{dc}{dz} = \frac{H}{K} c (1 - c) - \frac{1}{K} (I_1 - I_c),$$

(9)

where H/K and 1/K are the first, constant terms in the expansion of the coefficients of Eq. (3) into a Fourier series

$$\frac{H}{K} = \frac{1}{L} \int_0^L \frac{G\alpha}{\delta \left[\frac{1-\delta}{\delta\tau_x} G^2 + R_D + R_n \right]} dz,$$

(10)

$$\frac{1}{K} = \frac{1}{L} \int_0^L \frac{dz}{\delta \left[\frac{1-\delta}{\delta\tau_x} G^2 + R_D + R_n \right]}.$$

(11)

A calculation carried out with these formulas showed that at $R_n = 0$ the operating state of the column was completely determined by the two dimensionless parameters Pe_D and $S_0 = (Q_0/Lu_x) + 1$, where Q_0 is the additional amount of vapor introduced at the bottom of the apparatus.

Figure 2 shows curves illustrating the dependence of the value $HL/K\epsilon$, which is proportional to the logarithm of the separation coefficient of the column when there is no take-off, on these parameters (curves are given for the case $D_{12} = D_{13}$, $R_D = 0$, $R_n = 0$).

The same figure also shows as a dotted line the dependence of the value $H^2\text{Pe}_D/K\epsilon^2 u_x$, which is proportional to the separating capacity of the column, on Pe_D. Calculation shows that the separating capacity does not depend on S_0 and reaches a maximum value at $\text{Pe}_D = 2.82$.

Experimental Investigation of Apparatus

A large part of the experiments was performed with the neon isotopes $Ne^{20} - Ne^{22}$, due to the convenience of measuring them mass-spectrometrically. An investigation was made of the dependence of the separation coefficient under conditions of no take-off on the total amount of vapor introduced into the apparatus. As would be expected, with an increase in the vapor consumption, with a fixed value of S_0, the enrichment factor of the column, f, increased. In experiments with neon it reached a maximum value of 9. We were only prevented from obtaining a larger separation by the inadequacy of the heater power. Fig. 3 shows one of the curves obtained in these experiments.

An investigation was also made of the dependence of the enrichment factor of the column under conditions of no take-off and with constant vapor consumption on S_0 (Fig. 4). These experiments confirmed the possibility of controlling the operating state of the apparatus by changing the heater power of the evaporator and the amount of additional vapor introduced at the bottom of the column and that led off into the condenser at the top.

Similar investigations were also performed with carbon isotopes in the form of methane. The maximum enrichment factor of the column which could be obtained using xylene vapor as the accessory vapor did not exceed 1.9. We considered that the reason for the comparatively low enrichment was losses specific for diffusion

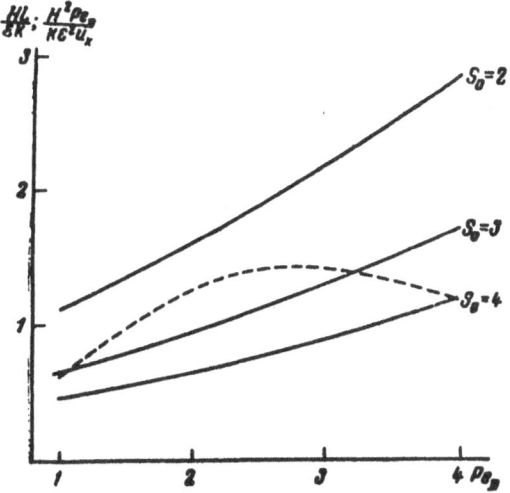

$\frac{HL}{\epsilon K}; \frac{H^2 Pe_D}{K\epsilon^2 u_x}$

Fig. 2. The dependence of the column parameters on the Peclet diffusion number (HL/εK — solid line and $H^2 Pe_D / K\epsilon^2 u_x$ — dotted line).

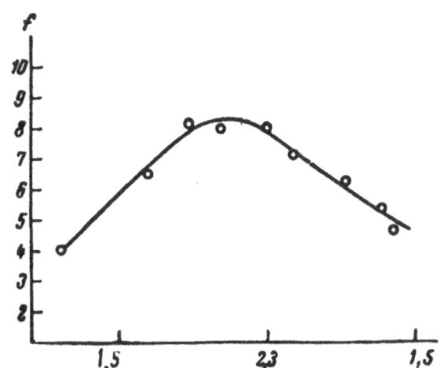

Fig. 4. Dependence of the enrichment factor of the column f on S_0 in the separation of neon isotopes, $Ne^{20} - Ne^{22}$ under conditions of no take-off ($u_x = 0.45$ cm/sec.

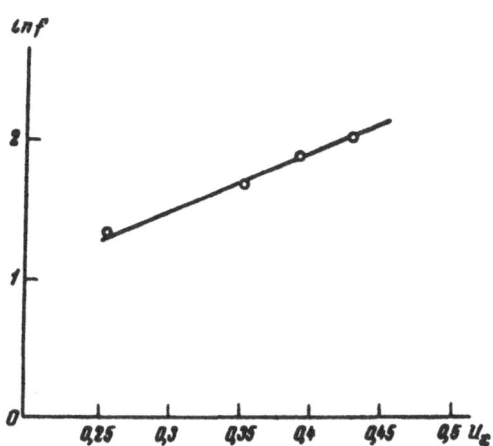

Fig. 3. Dependence of the logarithm of the enrichment factor of the column in the separation of neon isotopes ($Ne^{20} - Ne^{22}$) under conditions of no take-off on current density of the vapor through the diaphragm u_x (u_x is expressed in normal milliliters of vapor passing through 1 cm^2 of diaphragm per second).

columns caused by the solubility of the gas in the liquid films running down the condenser. By replacing the xylene by nitrobenzene, in which methane is very much less soluble, and simultaneously lowering the operating pressure to 50 mm Hg, we were able to increase the enrichment factor for $C^{12}H_4 - C^{13}H_4$ to 2.5.

Although S_0 and Pe_D do not depend on the pressure, when the pressure was increased above 200 mm, the enrichment factor rapidly fell. A possible reason for this is a decrease in the gas flow in the space between the diaphragm and the condenser with an increase in pressure, which becomes comparable with the rate of drainage of the liquid film down the condenser. This may lead to noticeable mixing as a result of entrainment of the adjacent gas layer by the film.

Measurements were made on the dependence of the neon separation on the take-off. According to the data of these experiments, and using known formulas for the case where $Pe_D \approx 3$ and $S_0 = 2.5$, we determined the parameters of the column H = $7.2 \cdot 10^{-6}$ g/cm · sec and K = $4.5 \cdot 10^{-4}$ g/sec, and also calculated the specific separation capacity $H^2/4K = 0.3 \cdot 10^{-7}$ g/cm^2 · sec. The fact that H and K were several factors smaller than would be expected from theory indicates the presence of factors in the apparatus, leading to a decrease in G. At present more careful measurements are being performed.

An experimental cascade was constructed from ten columns in series and isotopes of neon were obtained on this. Its capacity was 400 cm^3 of neon per day with an Ne^{22} content of about 99%. At present, the cascade has begun operation on the separation of carbon isotopes.

LITERATURE CITED

[1] H. Korsching, Z. Naturforsch. 6a, 213 (1951).

[2] M. Cichelli, W. D. Weatherford and Bowman, Chem. Eng. Progr. 47, 63, 111 (1951).

[3] D. Heyman and J. Kistemaker, J. Chem. Phys. 24, 165 (1956).

[4] K. Clusius and G. Dickel, Naturwiss. 26, 546 (1938).

[5] B. V. Deriagin, P. S. Prokhorov and A. D. Malkina, Collection: New Physicochemical   Methods of Investigating Surface Phenomena (Moscow, 1950) p. 155.

[6] C. F. Curtiss and J. O. Hirschfelder, J. Chem. Phys. 17, 550 (1949).

[7] M. Benedict and A. Boas, Chem. Eng. Progr. 47, 51 (1951).

A FRACTIONATING COLUMN FOR PREPARING BF$_3$ ENRICHED IN THE ISOTOPE B^{10}

Iu. K. Miulenfordt, G. G. Zivert and T. A. Gagua

For increasing the efficiency of counting equipment recording slow neutrons, a need has arisen for boron compounds, in which the concentration of the isotope B^{10} has been increased 3.5-4 times in comparison with the natural concentration. In a natural mixture, the B^{10} concentration is 18.6%. Thus, the problem of preparing boron compounds with a B^{10} content of approximately 80% has arisen.

A comparative evaluation of separation methods shows that the most acceptable process for the concentration of B^{10} is fractional distillation of boron trifluoride. It was established that B^{11}F$_3$ is more volatile than B^{10}F$_3$.

The separation coefficient, i.e., the ratio of the saturated vapor pressures of the isotopes in the liquid− −vapor equilibrium for BF$_3$ at −103° and a BF$_3$ pressure of 630 mm Hg, equals

$$\alpha = \frac{P_{\mathrm{B^{11}F_3}}}{P_{\mathrm{B^{10}F_3}}} = 1,0082 \, .$$

Such a value for the primary effect is quite acceptable for realization of a high-efficiency fractionation process. In our opinion, the use of BCl$_3$ is less profitable due to the lower separation coefficient ($\alpha = 1.0030$ at the boiling point), and some advantages, connected with the fact that the process is carried out at room temperature, probably cannot compensate for the loss in separation effect.

Since the working temperature of the process is in a low-temperature region ($\sim -100°$), the problem of achieving adiabatic conditions is complicated. To achieve adiabatic conditions, all the main sections of the fractionating column were connected to a cooling system, in which liquid ethylene at atmospheric pressure was circulated. The ethylene vapor condenser was connected through a thermal resistance to a vessel filled with liquid oxygen. As the problem of stabilizing the pressure in the ethylene system was successfully solved, adiabatic working conditions for the column were achieved.

The fact that B^{11}F$_3$ is more volatile than B^{10}F$_3$ determined the disposition of the main sections in the construction of the column; the still was placed at the top of the column; an evaporator was introduced below the column for complete return of the reflux; the B^{10} was taken off below the column.

Description of the Column

The essential plan of the column is shown in Fig. 1. The fractionating tube 4, which was 12 m long and had an internal diameter of 12 mm and a wall thickness of 0.25 mm, was made from cupronickel. The tube was packed with rings 1.2 mm in diameter, made from Constantan wire 0.25 mm in diameter. The values of the height of an equivalent theoretical plate (HETP) for a column with this packing and for various specific circulations are given in the table.

The bulk weight of the packing was 2.5 g/cm^3. The surface of the packing was 50.6 cm^2/cm^3. The fraction of free space was 0.7.

The fractionating tube was surrounded by a cylindrical vessel, through whose walls flowed liquid ethylene.

The walls of the vessel 6 and 7 were tubes of stainless steel, measuring 19 × 21 and 26 × 28 mm. At the bottom they were connected with a flexible metal bellows 5. The ethylene vessel was surrounded by a stainless steel shielding tube 8, measuring 35 × 37 mm.

109

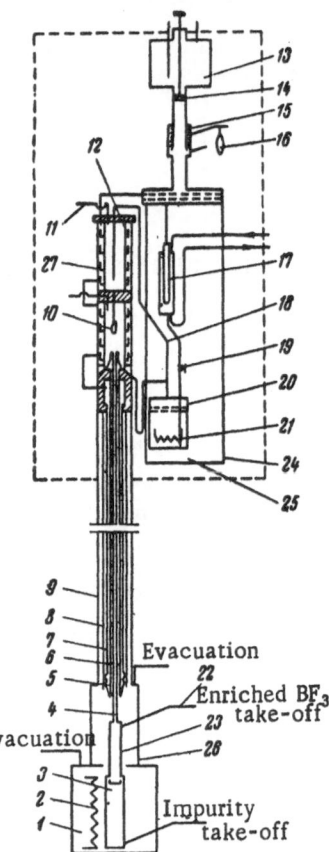

Fig. 1. Block plan of fractionating column.

TABLE

Specific circulation, $cm^3/cm^2 \cdot sec$	HETP, cm, for a column diameter of	
	7.6 mm	11 mm
24.8	2.2	2.3
15.2	2.0	2.1
8.3	1.8	1.9

The fractionating tube, the ethylene vessel surrounding it, and the shielding tube were placed in a vacuum jacket 9, which was a stainless steel tube 45 × 48 mm. The bottom of the jacket was connected with a flange to the vessel 26, which was necessary for convenience in assembling the lower part of the column. The jacket was pumped out to a pressure of ~10^{-4} mm Hg. To reduce the heat losses by thermal radiation, the inner walls of the vacuum jacket, the inner and outer walls of the shielding tube, and the outer wall of the ethylene vessel were polished to a mirror finish.

Beneath the fractionating tube was placed a small purification column 23, made from a cupronickel tube 12 × 12.5 mm and filled with the same packing as the main column. The purification column was used to purify the BF_3 from less volatile impurities such as HF and SiF_4.

Beneath the purification column was an evaporator 3, made of a copper tube 24 × 30 mm, and filled with rings with an internal diameter of 3 mm. The rings were made from German silver wire, 0.3 mm in diameter. The packing was necessary to retain solid SiF_4.

The evaporator was surrounded by vessel 1, which was filled with gaseous nitrogen when the column was operating. The evaporator received a certain amount of heat from the surrounding medium by heat transfer. If the evaporator needed more heat, the heater 2 was switched on. The power supplied to the evaporator was selected so that there was complete return of the reflux as vapor.

The enriched product was taken off through tube 22, placed between the column and the purifier. Impurities in the BF_3 (SiF_4 and HF) were taken off from the bottom of the evaporator.

The condenser consisted of two chambers, connected with a liquid seal 10. Cups filled with liquid ethylene were disposed around the walls of the condenser. BF_3 vapor from the still entered the upper chamber, condensed and drained through the liquid seal into the column.

In this way liquid BF_3 with the same isotope concentration as the boron in the still of the column was introduced into the upper part of the column.

The BF_3 vapor passing from the column, condensed in the lower condenser and drained into the still. The still 20, with a capacity of 700 cm^3, was placed at the top of the column. It was made from molybdenum glass, which made it possible to observe the boiling process and at the same time control the liquid level through a corresponding transparent plastic window. The still contained a heater 21, which consisted of a spiral of Nichrome wire.

The circulation in the column was kept stable by good thermal insulation of the still (it was placed in a special jacket 25, cooled with ethylene 24) and good stabilization of the power supplied to its heater.

The apparatus was provided with a system for renewing the contents of the still without disrupting the normal operating state of the column.

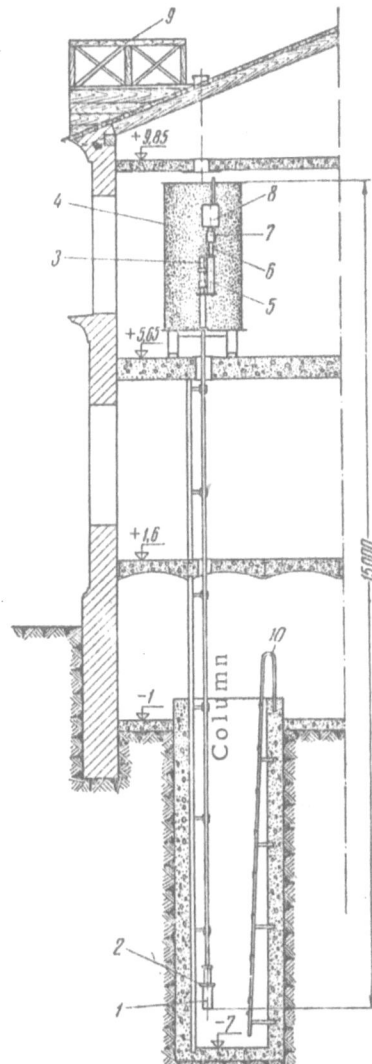

Fig. 2. Fractionating apparatus: 1) BF_3 evaporator; 2) take-off point for enriched product; 3) double condenser for BF_3; 4) thermal insulation; 5) screen around still; 6) column for ethylene purification; 7) ethylene condenser; 8) liquid air container; 9) platform for assembling column.

The BF_3 in the still was replaced in the following way: impoverished liquid BF_3 was withdrawn from the still through the valve 19, into the auxiliary condenser 17, through which it was removed from the system; fresh BF_3 was condensed in the auxiliary condenser and run into the still through the valve 19.

The condenser, still jacket, auxiliary condenser and ethylene vessel were cooled with a closed cooling system. The cooling substance was ethylene and its saturated vapor pressure was kept at 1 atm., i.e., the temperature was −103.7°.

The cooling system had its own condenser 15, which was connected by means of a piston 14, through a thermal resistance to vessel 13, filled with liquid oxygen. In this way the temperature of the ethylene condenser was regulated by changing the length of the thermal resistance (moving the piston).

To fill the cooling system, gaseous ethylene was passed into the ethylene condenser. Here it condensed and drained down, cooling in turn the auxiliary still, the jacket, the condenser and, finally, the ethylene vessel. The ethylene vapor passed from the cooling system into the ethylene condenser, condensed, and the ethylene again drained into the system. The ethylene vapor pressure in the cooling system was kept constant with a simple automatic regulator. This consisted of a spherical rubber bulb 16, attached to the upper part of the ethylene condenser. This bulb was filled with a gas that did not condense at the temperature of the ethylene condenser gas, for example hydrogen. The ethylene vapor flushed it into the upper part of the system and it filled the bulb and the upper part of the condenser. The ethylene vapor pressure was thus adjusted to approximately atmospheric. With an increase or a decrease in the pressure in the system there was a corresponding displacement of hydrogen from the condenser into the chamber, or the reverse. As a result of this the condensation limit was moved along the heat-conducting body and the pressure in the system was reduced. Variations in the atmospheric pressure were quite slow and did not essentially affect the operation of the column. For reliable thermal insulation, all the upper part and the liquid oxygen container were placed in a housing filled with granulated cork.

Positioning the fractionating column did not require special accommodation. The upper part of the column (Fig. 2), together with all the vacuum and electrical equipment, was housed in a room 20 m^2 and 4.1 m high, on the second floor. From here the column passed down through the first floor into a pit.

Basic Data

Two years' operation of the column described established the column parameters, the optimal process parameters, and the operating specifications.

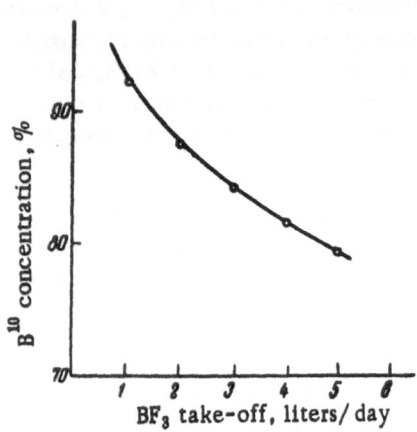

Fig. 3. Relation of B^{10} concentration in material taken off to take-off rate.

The column had 600 theoretical plates. After 20 days operation of the column without withdrawing material, an enrichment of 95-96% B^{10} was attained. The relation of the B^{10} concentration in the material taken off to the take-off rate is shown in Fig. 3. Since a B^{10} concentration of approximately 80% was required, the column was operated with a take-off of 4 liters per day. The optimal heating power for the still was 22.4 watts, which corresponded to a specific BF$_3$ circulation of $1.03 \cdot 10^{-3}$ mole $\cdot$ sec^{-1} $\cdot$ cm^{-2}.

Of the total liquid oxygen consumption (32 liters per day) approximately half was consumed in condensation of the BF$_3$ and half for condensation of the ethylene and heat losses. The B^{10} concentration in the still was allowed to fall to 15-16%. With the given still capacity (approximately 700 cm^3), the contents were renewed when 20 liters of enriched BF$_3$ had been removed.

The technological and operational data obtained are a basis for the design and construction of a high-efficiency column with a high concentrated B^{10} output.

AN INVESTIGATION OF THE SEPARATION OF THE STABLE ISOTOPES
OF LIGHT ELEMENTS

N. M. Zhavoronkov, O. V. Uvarov and S. I. Babkov

Scientific investigations on the separation of the stable isotopes of light elements and the study of their properties were begun in the USSR as early as 1936-1937 [1], and were developed in the post-war years. The present report presents the basic results achieved in the Mixture Separation Laboratory of the Karpov Physicochemical Institute* from investigations on the separation of oxygen, nitrogen, boron, carbon, and silicon isotopes.

The main characteristic for assessing the possibility of using this or that method of enriching isotopes is the value of the separation coefficient. In the case of distillation, this coefficient is the relative volatility of the components $-\alpha$.

The separation coefficient or the relative volatility (α) is determined by evaporating a large amount of the substance, composed of different isotopes, to a small residue, sufficient for performing analyses, and then calculating the value by Rayleigh's formula. In some cases, the method used involves evaporation in an apparatus whose separating power is equal to one theoretical stage (plate). With a large difference in isotopic composition, the differential method of measuring the difference in vapor pressure of the substances is used. Sometimes the separation apparatuses themselves can be used to determine the value of the separation coefficient quite successfully; in this case, the value of α is calculated from data on the maximum enrichment at a known number of theoretical separation stages or on the rate of enrichment in multistage separation apparatuses. These methods were used to determine the separation coefficients for $H_2O^{16} - H_2O^{18}$, $B^{10}Cl_3 - B^{11}Cl_3$, $C^{12} - C^{13}$, for methane, ethane, and ethylene, and some other systems over a wide range of temperatures.

These investigations made it possible to choose the most efficient separation method. In a series of cases, the most convenient methods of separating stable isotopes of light elements practically were fractional distillation and chemical isotope exchange.

Work was carried out on investigating and choosing an efficient apparatus design and establishing optimal process conditions. Of all the types of separating apparatus proposed, the simplest was a column filled with a fine packing. For processes where a high hydraulic resistance was not permissible, the recommended apparatuses had a regular, completely wetted packing, for example, a type of horizontal column with slowly rotating discs, partially submerged in the liquid.** The height equivalent to one theoretical separation stage or theoretical plate (HETP) should not exceed several centimeters as for the separation of components with close physicochemical properties such as isotopes, hundreds, and even thousands, of theoretical separation stages are required. Although high-efficiency packing materials for low-capacity columns have already been proposed by Thode [2], Dixon [3] and Fenske [4] et al., in our work we also paid attention to the selection of efficient and cheap packings.

The most efficient packing consisted of steel wire spirals, formed by winding on conical, triangular needles. These packings were proposed by Levin [5] for the separation of mixtures in laboratory columns, and we studied their application to isotope separation in detail. These spirals may be prepared in large amounts and quite quickly on automatic and semiautomatic lathes.

---

*V. A. Maliusov, N. A. Malofeev, N. N. Seriugova, V. Iu. Orlov, G. N. Chernykh, V. A. Sokol'skii and G. A. Iagodin also participated in investigations connected with the study and development of methods of separating isotopes.

**A design of this type of apparatus was developed for the separation of isotopes and has been described in published papers.

In columns with a very large number of separation stages and a low capacity, particular attention must be paid to the creation of conditions close to adiabatic along the whole length of the column. Dostrovsky [6] has done a great deal in this field. Our experiments showed that in columns of up to 2 m high this could be achieved quite satisfactorily by the use of thermal insulation with electrical heating or with vapor jackets. For columns of more than 2-3 m high, it was best to use a high vacuum jacket, surrounded by a thermal insulator. For processes at elevated temperatures, a thermal insulation jacket with electrical heating was recommended. As is known, the problems of maintaining constant flows with time, and reducing the product losses to low values are no less important. High enrichments may be obtained only by observing these requirements.

The separation coefficient values for isotopes of the light elements differ little from unity in most cases. Therefore, as has already been indicated, a large number of separation stages is required to obtain significant enrichments. In these cases the problem of the rate of accumulation of one of the components is of particularly great importance since the time required to reach a stationary state is measured not in hours but in weeks and even months.

To control the direction of the concentration process and also to plan calculations it is necessary to know the theoretical rules of the concentration changes through the stages of the column with time. Existing methods of determining the rate of change of the mixture components at the column stages are approximate and give low values for the time to reach a given concentration, which differ substantially from the practical values. The quantiative deviation is particularly great at substantial enrichments, as the process approaches a stationary state. The method of Cohen [7] is an exception, but it is very complicated and inconvenient for practical use. We therefore undertook an investigation of the kinetics of the multistage separation of stable isotopes to develop a simple and sufficiently accurate method of determining the rate at which a process approaches a stationary state. In deriving theoretical equations for the rate of approach to a stationary state [8], i.e., the rate of change of concentration in a column for actual separation processes occurring according to an open scheme with total reflux, we started from the following premises.

1) At the very beginning of the separation process the concentration of the heavy component in the gas phase in any section of the column is above the equilibrium value and therefore mass exchange begins simultaneously in all sections.

2) The concentration of the heavy component at any K-th stage of the column increases, beginning from the original concentration N_0 and, simultaneously at all the stages of the column, it reaches the practically limiting value, asymptotically approaching the value $\overline{N}_k = \dfrac{N_0 \alpha^k}{1 + N_0(\alpha^k - 1)}$, which characterizes the stationary state.

Starting from these premises, one can assume with sufficient accuracy for calculation purposes that a similarity between the curves of concentration change through the stages of the column is maintained during all of the approach to the stationary state, and this is expressed by the relation

$$\frac{N_k - N_0}{N_1 - N_0} = \frac{\overline{N}_k - N_0}{\overline{N}_1 - N_0}.$$

Then the concentrations in the equation for the material balance of the column

$$L\left[N_0 - \frac{N_1}{\alpha - N_1(\alpha - 1)}\right] dt = H dN_1 + \ldots + H dN_k + \ldots + H dN_n + v_0 dN_n \quad (1) \qquad (1)$$

may be expressed in terms of the concentration in the upper stage, N_1.

Subsequent integration of (1) leads to an equation for the rate of approach to the stationary state:

$$t = \left[\frac{v_0}{L} \frac{\alpha^n - 1}{1 + N_0(\alpha^n - 1)} + \frac{H}{L} \sum_{1}^{n} \frac{\alpha^k - 1}{1 + N_0(\alpha^k - 1)}\right] \frac{1}{1 + N_0(\alpha - 1)} \times$$

$$\times \left\{\frac{\alpha - 1}{\alpha^n - 1}[1 + N_0(\alpha^n - 1)](N_n - N_0) + \frac{\alpha}{\alpha - 1} \ln \frac{(\alpha^n - 1)(1 - N_0)}{\alpha^n - [1 + N_0(\alpha^n - 1)]\frac{N_n}{N_0}}\right\}. \qquad (2)$$

114

In a series of cases for the separation of isotopes the limiting concentrations $\overline{N}_k = \dfrac{N_0 \alpha^k}{1 + N_0(\alpha^k - 1)}$, obtained in a single column, are small and may be expressed by the value $N_0 \alpha^k$ without appreciable error. Then Eq. (2) is converted into the equation

$$t = \frac{1}{L}\frac{\alpha}{\alpha - 1}\left[\left(H\frac{\alpha}{\alpha - 1} + v_0\right)(\alpha^n - 1) - nH\right]\ln\frac{\alpha^n - 1}{\alpha^n - \dfrac{N_n}{N_0}}. \tag{3}$$

In these equations: t — time in hours; v_0 — amount of mixture in the still (or evaporating system), in moles; H — "retention" in one separation stage (amount of mixture in a part of the column, equivalent to one theoretical plate), in moles; N_0 — concentration of the heavy component in the original mixture, in fractions of a mole; N_1, N_k and N_n — concentrations of the heavy component at the corresponding stages; $\overline{N}_1$, $\overline{N}_k$ and $\overline{N}_n$ — limiting concentrations (at the stationary state) of the heavy component at the corresponding stages; L — "loading" — the amount of mixture passing through the column in one direction, in moles per hour; n — number of theoretical separation stages in the column.

Using Eqs. (2) and (3) one may not only calculate the time to reach a given isotope concentration but, with experimental data on the course of enrichment with time available, one can also estimate the number of theoretical separation stages, as we did for experiments on the separation of oxygen and boron isotopes. The equations may also be used for the calculation or approximate evaluation of α in the separation of a mixture for which it is not known.

So as to test the equations derived and to compare them with experimental data, numerous experiments were carried out to study the change in concentration of nitrogen isotopes with time, using the chemical exchange between gaseous ammonia and the ammonium ion in ammonium nitrate solutions [8]. The experiments were performed in a horizontal column with rotating discs with constant "retention," using different "loadings" and removing samples for analysis at several points along the column at various times from the beginning of the enrichment process.

The values of the enrichment N_n/N_0 with time, calculated by Eq. (3), agreed well with experimental data. This was also confirmed by our experiments on the separation of oxygen and boron isotopes by fractional distillation, by the data of North and White [9] on the concentration of S^{34} by isotope exchange, and also by the experimental data of Berg and James [10] on the fractional distillation of methylcyclohexane–n-heptane mixture.

As a result of the investigations carried out and the solution of a series of engineering problems, and considering the work carried out in other countries, convenient methods were developed for preparing concentrates of rare isotopes of some light elements.

Preparation of the Heavy Oxygen Isotope, O^{18}

The preparation of H_2O^{18} concentrates may be realized by several methods, but some of them are of low capacity (thermal diffusion and diffusion), and others require the consumption of expensive chemicals (chemical exchange). The most convenient and simplest method is the fractional distillation of water.

To evaluate the operation and calculate the required height of a separating column, it is necessary to know the separation coefficients (relative volatility) of the $H_2O^{16} - H_2O^{18}$ system and their dependence on temperature. The published data of Wahl and Urey [11] and of Riesenfeld and Chang [12] differ considerably in the numerical values. Therefore we carried out an investigation to determine α.

The determination was achieved by two methods: 1) evaporation until equilibrium was established between vapor and liquid in an Othmer-type of apparatus, and 2) evaporation of a large volume of water to a very small residue with subsequent calculation by Rayleigh's formula. Both apparatuses were first calibrated with the $D_2O - H_2O$ system, for which the value of α is known. All determination of the separation coefficient for the $H_2O^{16} - H_2O^{18}$ system were performed with water previously enriched in H_2O^{18} (1.1-1.2 mole%) and with a normal deuterium oxide content (approximately 0.016 mole%). The measurements were performed in the temperature range from 15 to 100° and the separation coefficient changed from 1.0093 (15°) to 1.0032 (100°). The temperature dependence may be expressed by the equation

$$\log \alpha = 3.4492 \frac{1}{T} - 0.00781.$$

The description of the method and the results of the investigation were presented in [13]. The main difficulty in the separation of oxygen isotopes by the fractional distillation of water results from the low value of the separation coefficient. The preparation of concentrates with an H_2O^{18} content of about 50% requires a high-efficiency fractionation column with more than two thousand theoretical separation stages. The high separating capacity of the columns was achieved both as a result of their height and the use of an efficient packing of small spirals, made from steel wire 0.2 mm thick. The use of a cascade of columns with decreasing diameter is known to decrease the time required to reach a stationary state. However, even under these conditions, steady, continuous operation for several months is required.

After numerous experiments on the fractionation of water in glass columns 30 mm in diameter and about 2 m high, an apparatus was constructed from three successive metal columns 60, 52 and 20 mm in diameter and with a packing height of 16.5, 9.5 and 16.5 m. The first column was filled with a packing of spirals measuring 2.6 × 2.9 mm and for the second and third columns, the packing measured 1.6 × 1.9 mm. The construction of one of these columns was described in [14].

The columns could operate both in cascade and individually. The first column (with the biggest diameter) had an upper reservoir of 352 liters capacity, operating on stripping. During operation of the cascade, the concentrate had to appear in the still of the third column. We performed prolonged experiments to study the concentration process when the column was operated as a closed system. Here the concentrate obtained in the first and third columns was used for periodically supplying the second column and was added to the small (18-liter) upper stripping reservoir. After 2877 hours of continuous operation, the second column had a still concentration of H_2O^{18} equal to 24.5%; we also obtained other portions of water with from 4 to 21% of H_2O^{18}, totalling about 2 kg.

A study was made of the operation of all the columns and the essential technological parameters were measured. It was established that columns with small diameters (20 mm and less) were unprofitable; although the "retention" of these columns was low, the relative heat losses were very great, even with good thermostating, and this led to condensation of the vapor and, consequently, to a fall in mass transfer.

The most efficient column had a diameter of 52 mm and its charge reached 125 g/cm^2 · hour in our experiments. On this column the height equivalent to one theoretical stage reached about 0.6-1 cm, which corresponded to a maximum number of theoretical separation stages of about 1500. The investigations were performed at atmospheric and at reduced pressure.

The preparation of heavy water concentrates by chemical isotope exchange in the system $CO_2 - H_2O$, which we also studied, seemed less efficient to us. At the present time we are continuing work on studying the physicochemical properties of H_2O^{18} and H_2O^{17}.

Preparation of the Heavy Isotope N^{15}

The best of the methods of separating the stable isotopes of nitrogen, known at the present time, is that involving the chemical exchange between gaseous ammonia and the NH_4^+ ion in aqueous ammonium nitrate solution [15-19].

As with any multistage separation process, the concentration of the isotope N^{15} requires extremely rigid technological conditions. Therefore, the number of simultaneously regulated flows in the apparatus must be reduced to a minimum. The construction of the separating apparatus must ensure that the equipment can be brought into operation rapidly after unavoidable shutdowns and that a concentrate can be accumulated in the column without loss.

Investigations on the concentration of the heavy nitrogen isotope in high-capacity columns and horizontal apparatuses with rotating plane-parallel packings, which we performed, showed that a combination of these columns made it possible to realize a large-scale apparatus for the production of highly concentrated N^{15} with the minimum number of cascades (two) and feed pumps (two).

The plan of such an apparatus for the production of 10 kg of ammonium nitrate per year with a 50% N^{15} content in the ammonium nitrogen, and also the construction of the apparatus are described in our paper [20]. The first example of this apparatus for preparing concentrates of ammonium nitrate containing 10-20% N^{15} was constructed in one of our workshops and tested for a long period.

Recently, communications have appeared in the literature on the preparation of N^{15} concentrates by chemical exchange between nitrogen oxides and nitric acid, etc. [21]. Similar work is at present proceeding in our laboratory.

Preparation of the Light Isotope of Boron B^{10}

To prepare substances with a high B^{10} content one can use several methods of separating boron isotopes — — thermal diffusion, chemical isotope exchange, and fractional distillation. The latter method is the simplest and has the highest capacity. It has to be applied to liquid boron compounds such as the halides BF_3 (b.p. −101.7°), BCl_3 (b.p. 12.7°), and BBr_3 (b.p. +91.7°).

Urey [21] attempted to calculate the separation coefficient for the system $B^{10}Cl_3 - B^{11}Cl_3$ by statistical thermodynamics and obtained the value 1.013 at 25°.

By fractionation of BCl_3 on a glass laboratory column (2 m high and 20 mm in diameter with a packing of rings measuring 1.5 × 1.5 mm), Green and Martin [22] established that $B^{11}Cl_3$ was the more volatile component and not $B^{10}Cl_3$.

According to their data, the separation coefficient α for the normal boiling point was 1.0018.

Kats, Kukavadze and Serdiuk [23] carried out similar experiments on a laboratory column two meters high and 20 mm in diameter, and obtained a maximum enrichment factor of 1.4 on $B^{10}Cl_3$. Assuming that the column had 80 theoretical separation stages (according to their data on the fractionation of heavy water), they calculated that the separation coefficient was 1.0043 at the normal boiling point.

In our laboratory we determined the value of the separation coefficient (α) for the system $B^{10}Cl_3 - B^{11}Cl_3$ and its dependence on temperature by Rayleigh distillation. Since the separation coefficient is small, the ratio of the volumes before and after distillation was large in the experiments (not less than 5000). The distillation was performed in two stages; the final residue was not more than 0.5 g. The isotopic analysis of the residue was performed on an MS-4 mass spectrometer. The experiments were carried out in the temperature range from +12.7 to −85°.

The separation coefficient at the normal boiling point was 1.003. With a decrease in temperature the value fell sharply. At −61.7° the vapor pressures of $B^{10}Cl_3$ and $B^{11}Cl_3$ were equal, and at lower temperatures $B^{10}Cl_3$ was the more volatile. The dependence of the separation coefficient on temperature could be expressed by the equation

$$\log \alpha = 0.00483 - \frac{1.00757}{T} .$$

A description of the procedures and the results of determining the separation coefficients of boron isotopes by equilibrium evaporation of BCl_3 are presented in our paper [24].

We also performed experiments on preparing $B^{10}Cl_3$ concentrates by fractional distillation of BCl_3 in a stainless steel column 12 m high and 21 mm in diameter with a packing of wire spirals measuring 2 × 1.5 mm. The column had a condenser and a 5-liter stripping vessel above, which was always filled with liquid, and a 100 cm³ still below in the form of a double-walled cylinder. An electrical heater was placed inside the cylinder. The column was thermally insulated with a vacuum jacket.

The first series of experiments were performed at atmospheric pressure (boiling point +12.7°) with the condenser cooled with solid carbon dioxide and acetone. The column was operated continuously for 42 days with a load of 600 cm³/hour. The column did not work sufficiently steadily. The increase in the $B^{10}Cl_3$ enrichment varied with time. As a result of the experiment it was possible to reach a 5.5-fold enrichment and to obtain 100 g of product with a 57% $B^{10}Cl_3$ concentration.

The second series of experiments was performed with an elevated pressure in the column and better thermostating conditions. In this series of experiments a five-fold enrichment was obtained after 23 days continuous operation.

Experiments on the preparation of $B^{10}Cl_3$ concentrates at an elevated pressure are being continued. However, results already obtained have shown that the separation of boron isotopes by fractionation of BCl_3 may be realized considerably more simply than by the fractionation of BF_3. Although the separation coefficient for the system $B^{10}F_3 - B^{11}F_3$ at the normal boiling point is 1.009, i.e., a factor three greater than that for the system $B^{10}Cl_3 - B^{11}Cl_3$, the fractionation of BF_3 at a temperature of $-101°$ requires very good thermostating and also the use of liquid air as the coolant for the condenser, which considerably complicates the construction and operation of the apparatus [25].

Investigation of the Separation of the Stable Isotopes of Carbon and Silicon

For some purposes the stable isotope C^{13} is required, as well as the radioactive C^{14}. Urey et al. [26], described a method of concentrating C^{13} by chemical exchange.

The best method of obtaining C^{13} is apparently the fractional distillation of CO and CH_4 in packed columns. In the work of Deviatykh and Zorin [27] a study was made of the separation conditions and the preparation of $C^{13}O$ concentrates. To explain some rules on the physicochemical behavior of molecules with different isotopic compositions it is interesting to determine the separation coefficient for various substances of similar chemical structures.

In connection with this, we determined the separation coefficient of carbon isotopes for simple hydrocarbons — ethylene, ethane, and methane. An investigation was made of the distribution of heavy carbon in the equilibrium phases, liquid–vapor, in relation to temperature, by means of Rayleigh distillation. This method imposes extremely drastic requirements as regards the purity of the product. Therefore, carefully purified starting materials were used. The ratio of the initial and the final volumes in the experiments on equilibrium evaporation was 10,000.

The separation coefficient of carbon isotopes for ethylene had a very low value. Thus, at the normal boiling point it equalled 1.0019 ± 0.0004 and increased insignificantly with an increase in temperature.

The separation coefficient of carbon isotopes for ethane depended considerably on temperature and changed from 1.000 ± 0.0003 at 184.5° to 1.0022 at 130°K, i.e., decreased with an increase in temperature. The separation coefficient for the system $C^{12}H_4 - C^{13}H_4$ was determined previously by Groth [28] and then by Deviatykh and Zorin [27], but the data they obtained differed considerably. Thus, at the normal boiling point $\alpha = 1.0049$ according to Groth, and 1.0114 according to the data of Deviatykh and Zorin. The results we obtained were close to the data of Deviatykh and Zorin and changed from 1.0125 (at 90.55°K) to 1.0068 (at 111.8°K).

A comparison of the separation coefficient of carbon isotopes at the liquid–vapor equilibrium for methane and for ethane at the triple point indicates that α will decrease with an increase in the number of carbon atoms for other saturated hydrocarbons. A description of the procedure and results of the investigation was presented in [29].

Up to the present time, the isotopes of silicon have only been separated in extremely small amounts by the electromagnetic method. We carried out an experiment on the fractional distillation of silicon tetrachloride, the most readily available and convenient to handle of the halogen compounds of silicon.

In analogy with the fractionation of boron trichloride and carbon tetrachloride, one would expect that the heavy isotope of silicon would concentrate in the vapor phase. The fractionation of silicon tetrachloride in a glass column with about 100 theoretical separation stages confirmed this idea. However, the enrichment achieved was very insignificant and did not make it possible to calculate the separation coefficient even approximately [30].

As a result of the investigations carried out both in our and in other laboratories, at the present time we have at our disposal many methods of separating isotopes of the light elements, which are of great importance for practical and scientific purposes. However, much work still remains to be done to develop new and more efficient methods of preparing highly enriched concentrates of rare isotopes and to study their physicochemical properties.

LITERATURE CITED

[1] A. I. Brodskii, Prog. Chem. 6, 152 (1937); Prog. Phys. Sci. 20, 153 (1938); J. Appl. Chem. 13, 663 (1949); Chemistry of Isotopes (Acad. Sci. SSSR Press, 1952); A. I. Brodskii and O. K. Skarre, J. Phys. Chem. 18, 453 (1939).

[2] H. G. Thode, S. R. Smith and F. O. Walking, Can. J. Res. B22, 127 (1944); H. G. Thode and F. O. Walking, Can. J. Res., B. 20, 61 (1942).

[3] Dixon, J. Soc. Chem. Ind. 68, 88 (1949).

[4] M. R. Fenske, Ind. Eng. Chem. An. Ed. 17, 580 (1949).

[5] A. I. Levin, Petroleum Economy No. 10, 47 (1949).

[6] I. Dostrovsky, E. D. Hughes and D. R. Lewellyn, Nature 161, 9100, 858 (1948); I. Dostrovsky, D. R. Lewellyn, B. H. Vromen, J. Chem. Soc. 9, 3509 (1952).

[7] K. Cohen, J. Chem. Phys. 8, 583 (1940).

[8] S. I. Babkov and N. M. Zhavoronkov, Proc. Acad. Sci. USSR 106, 5, 877 (1956).

[9] E. North and R. White, Ind. Eng. Chem. No. 10 (1951).

[10] C. Berg and J. James, Chem. Eng. 307 (1948).

[11] M.H. Wahl and H. C. Urey, J. Chem. Phys. 1, 411 (1933).

[12] E. H. Riesenfeld and F. Z. Chang, Z. Phys. Chemie 33, 120 (1936).

[13] N. M. Zhavoronkov, O. V. Uvarov and I. I. Seriugova, Coll.: The Use of Tracers in Analytical Chemistry (AN SSSR, 1955).

[14] O. V. Uvarov, V. A. Sokol'skii and N. M. Zhavoronkov, Chem. Ind. No. 7, 20, 404 (1956).

[15] J. Kirschenbaum, J. Chem. Phys. No. 7, 44 (1947).

[16] H. G. Thode and H. C. Urey, J. Chem. Phys. No. 7, 34 (1939).

[17] E. W. Becker and H. Baumgärtel, Z. Naturforsch. 1, 119, 514 (1946).

[18] A. Sugimoto, R. Nakane and T. Watanabe, Bull. Chem. Soc. Japan 24, 153 (1951).

[19] V. V. Ottesen and M. E. Aerov, J. Phys. Chem. (USSR) 30, 6, 1356 (1956).

[20] S. I. Babkov and N. M. Zhavoronkov, Chem. Ind. No. 7, 4, 388 (1955).

[21] H. C. Urey, Chemistry of Isotopes, Coll. 1 [Russian translation] (IL, 1956) p. 86.

[22] M. Green and G. Martin, Trans. Faraday Soc. 48, 5 (1952).

[23] M. Ia. Kats, G. M. Kukavadze and R. L. Serdiuk, J. Tech. Phys. (USSR) 26, 12, 2744 (1956).

[24] I. I. Sevriugova, O. V. Uvarov and N. M. Zhavoronkov, J. Atomic Energy (USSR) No. 4, 113-117 (1956).

[25] Iu. K. Miulenfordt, G. G. Zivert and T. A. Gagua, Report to the All-Union Conf. on the Use of Radioactive and Stable Isotopes and Radiation in the National Economy and Science, April 4-12, 1957, Moscow.

[26] J. Roberts, H. G. Thode and H. C. Urey, J. Chem. Phys. 7, 137 (1939); C. A. Hutchinson, D. W. Stedwart and H. C. Urey, J. Chem. Phys. 8, 532 (1940); A. F. Reid and H. C. Urey, J. Chem. Phys. 11, 403 (1943).

[27] G. G. Deviatykh and A. D. Zorin, J. Phys. Chem. (USSR) 30, 5, 1133 (1956).

[28] W. Groth, H. Ihle and A. Murrenhoff, Z. Naturforsch. 9, Bd. 9-a, 805-806 (1955).

[29] G. A. Iagodin, O. V. Uvarov and N. M. Zhavoronkov, Proc. Acad. Sci. USSR 111, 2, 384 (1956).

[30] V. Iu. Orlov and N. M. Zhavoronkov, J. Appl. Chem. 29, 959 (1956).

THE SEPARATION OF CARBON ISOTOPES

N. N. Tunitskii, G. G. Deviatykh, M. V. Tikhomirov,

A. D. Zorin and N. I. Nikolaev

For carbon, as for other light elements, physicochemical methods are most advantageous for the separation of the isotopes. In 1940 [1], a method was proposed based on the use of the isotope exchange reaction between hydrogen cyanide and the cyanide ion in aqueous solution. This method was extremely effective. Thus, on a column 19.5 m high, a 25-fold enrichment of carbon in the isotope C^{13} was obtained. However, the tendency of hydrogen cyanide to polymerize, and the toxicity of it and its salts, make the cyanide method extremely unattractive. In 1943 a second method was proposed using the isotope exchange reaction between carbon dioxide and the bicarbonate ion in aqueous solution [2]. The disadvantage of the bicarbonate method was the low reaction rate of the isotope exchange with all the packings investigated, and this made it necessary to construct bulky apparatuses; also, too long a time was required for the apparatus to reach operating conditions. It was suggested that the exchange reaction between carbon dioxide and the bicarbonate ion could be accelerated by catalytic packings [2].

We devised experiments to study the effect of the nature of the packing on the separation factor. A series of packings, used in [2], were investigated. The results of our experiments [3] and the data of other authors (Table 1) may be explained starting from the hypothesis that the packings have no catalyzing action in all these experiments. The slowest stage of the exchange is the hydration of the carbon dioxide, which occurs by two parallel reactions [4,5]:

$$CO_2 + OH^- \underset{\longleftarrow}{\overset{K_1}{\longrightarrow}} HCO_3^-;$$
$$CO_2 + H_2O \underset{\longleftarrow}{\overset{K_2}{\longrightarrow}} H_2CO_3.$$

The equation for the maximum separation coefficient of the column when no material is withdrawn will have the following form in this case:

$$\ln F = \frac{1}{\alpha} (K_1 [OH^-] + K_2 [H_2O]) (\alpha - 1) \frac{c}{\omega L} z, \tag{1}$$

where F is the separation coefficient of the column; α is the separation coefficient; c is the solubility of carbon dioxide under the given conditions; L is the rate of flow of the bicarbonate, mole/$cm^2 \cdot$ min; z is the length of the column; and ω is the ratio between the total volume of the column and the volume occupied by the liquid.

As Table 1 shows, there is good agreement between the values of the height of an equivalent theoretical plate (HETP), obtained by experiment and calculated from Eq. (1).

Experiments at elevated pressure on a two-meter column, whose results are illustrated in Fig. 1, showed that there is an essentially linear relation between the number of theoretical plates in the column n and the carbon dioxide pressure p, as follows from Eq. (1). By increasing the pressure, the height of a theoretical plate may be reduced to a value of a few centimeters, which is determined by the permeability of the intersurface layer and not by the rate of the hydration reaction. However, at the same time, the separation coefficient is also reduced. For example, at a pressure of 30 atm the value of $\alpha - 1$ will equal only half of its value at a pressure of 1 atm. In addition, when the pressure is increased, the gas retention of the column and the lower

TABLE 1

Effect of the Type of Packing on the Separation Coefficient of the Column (p = 1 atm, T = 20°)

Packing	L 10^{+3} mole per cm^2·min	$\frac{1}{\omega}$	HETP, exp. value, cm	HETP, calc. from [1], cm	HETP· $\frac{1}{\omega}$	Source
Activated birchwood charcoal	2.0	0.59	68	61	33.3	Present work
AG grade activated charcoal	2.0	0.55	66	64	29.6	The same
Granulated aluminum oxide	2.0	0.62	62	59	32.2	" "
Stainless steel wire spirals	2.0	0.17	157	168	26.7	" "
Stainless steel wire spirals covered with carbon	2.0	0.17	157	168	26.7	" "
Nichrome wire spirals	2.0	0.17	150	168	25.5	" "
Chamotte	—	—	121	—	—	Data of Zhavoronkov et al.
Chamotte + 3% Fe_2O_3	—	—	111	—	—	The same
Chamotte + 3% Ni_2O_3	—	—	108	—	—	" "
Chamotte + 3% Al_2O_3	—	—	76	—	—	" "
Chamotte + 10% Al_2O_3	—	—	141	—	—	" "

TABLE 2

Relative Vapor Pressure of Heavy-Carbon Methane

T, °K	$\dfrac{PC^{12}H_4}{PC^{13}H_4}$
90.5 (normal melting point)	—
91.44	1.0114 ± 0.0006
97.55	1.0105 ± 0.0003
103.1	1.0111 ± 0.0007
111.8 (normal boiling point)	1.0098 ± 0.0003

TABLE 3

Relative Vapor Pressure of Isotopic Molecules of Carbon Monoxide

T, °K	$\dfrac{PC^{12}O}{PC^{13}O}$	$\dfrac{PCO^{16}}{PCO^{18}}$
68.1 (normal melting point)	—	—
71.3	1.0120 ± 0.0009	1.0154 ± 0.0019
78.3	1.0105 ± 0.0017	1.0088 ± 0.0012
81.8 (normal boiling point)	—	—
82.1	1.0108 ± 0.001	1.0043 ± 0.0018

reservoir is increased and this increases the time for the apparatus to reach operating conditions. With a column 16 m long, operating at 4 atm pressure, the rate of accumulation of the isotope C^{13} at the bottom end was only 2% per day of the original concentration. The rate of accumulation of the isotope was practically independent of reflux density.

After we had arrived at the conclusion that the bicarbonate method could hardly be improved substantially, at the end of 1952, investigations were begun on the possibility of separating carbon isotopes by fractional

TABLE 4

Isotopic molecules	Triple point temperature, °K	p'/p", calculated values*	p'/p", experimental values*	Source
$C^{12}H_4 - C^{13}H_4$	90,5	1,0072	1,0048	[9]
			1,011	[6]
$CH_4 - CD_4$	90,5	0,9958	0,9872	[11]
$H_2O - D_2O$	273,2	1,18	1,255	[12]
$H_2O^{16} - H_2O^{18}$	273,2	1,012	1,011	
$C^{12}O - C^{13}O$	68,1	1,012	1,012	Our measurements
$CO^{16} - CO^{18}$	68,1	1,016	1,019	the same

*p' and p" are the pressures of the light and heavy isotopic components, respectively.

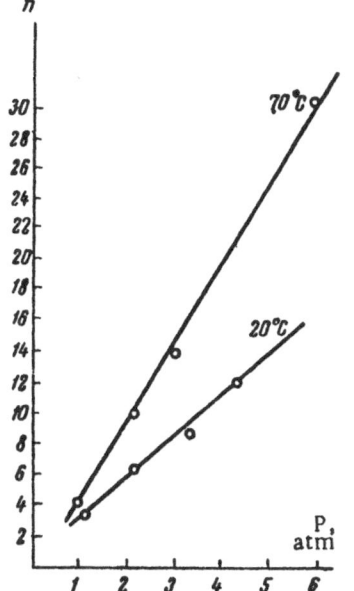

Fig. 1. Relation of the number of theoretical plates of a two-meter bicarbonate column to the pressure.

distillation of methane and carbon monoxide since these substances have the greatest relative differences in the masses of the isotopic molecules. The separation coefficients were determined by Rayleigh distillation [6]. The results of the determinations are given in Tables 2 and 3.

Of all the carbon compounds studied, methane and carbon monoxide gave the largest separation coefficients for carbon isotopes [7]. The results for carbon monoxide agreed well with the results of [8]. The results for methane were subsequently confirmed by [7]. Noticeably different data for the separation coefficient of isotopic molecules of methane were obtained in [9,10]. It is interesting that in the one case [9] the measurements were performed by the differential method using approximately 100% $C^{13}H_4$. A value of α of 1.001 was obtained. In the other work [10], the measurements were also performed by the differential method, but using 10% $C^{13}H_4$, and $\alpha = 1.0049$ was obtained. In [7] the measurements were performed by Rayleigh distillation.

An attempt was made to calculate the separation coefficient theoretically. In the calculation the assumptions were made that the liquid had a quasicrystalline structure and the chemical potential of a component in the liquid phase could be expressed by a Debye function. In addition, it was assumed that the van der Waal's forces did not depend on the isotopic composition of the molecule. The results obtained for a series of substances at the triple point temperature were in satisfactory agreement with experimental data, as is shown by Table 4.

Experiments on the distillation of carbon monoxide were performed on metal columns 3.75 and 12 m high and a glass column 1 m high. The column 12 m high had a two-meter stripping section. The plan of one type of metal column is illustrated in Fig. 2. The column was a stainless steel tube, whose outer surface was polished to reduce radiant heat exchange. The upper end of the column was connected to a condenser of pure copper, containing 9 liters of liquid nitrogen. The condenser had a 250 cm³ capacity reservoir for the distilled material. The lower part of the column ended in a glass still with dropping tubes and a heating element. The metal and glass were joined by means of a kovar tube. The column and the condenser were placed in a vacuum jacket, whose inner wall was shielded with sheets of polished tinplate. The vacuum in the jacket was maintained with two continuous diffusion pumps and equalled ~ 2 · 10^{-6} mm Hg. The packing consisted of spirals of stainless steel wire with $\alpha \sim 0.2$ mm; the dimensions of the spirals were 1.5 × 1.5 mm.

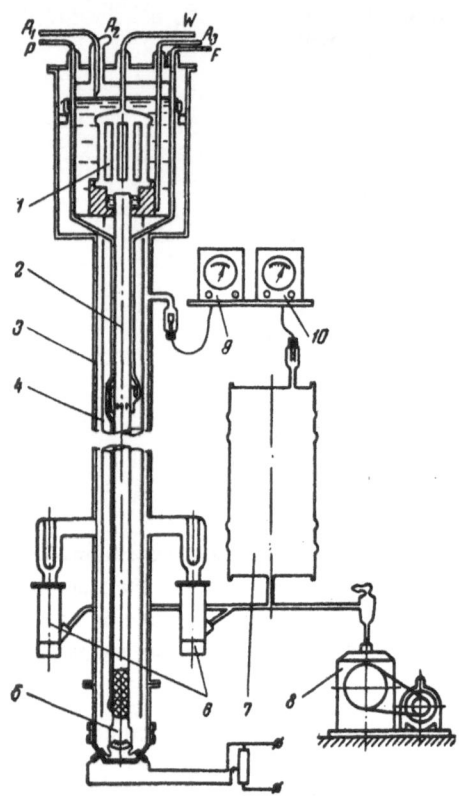

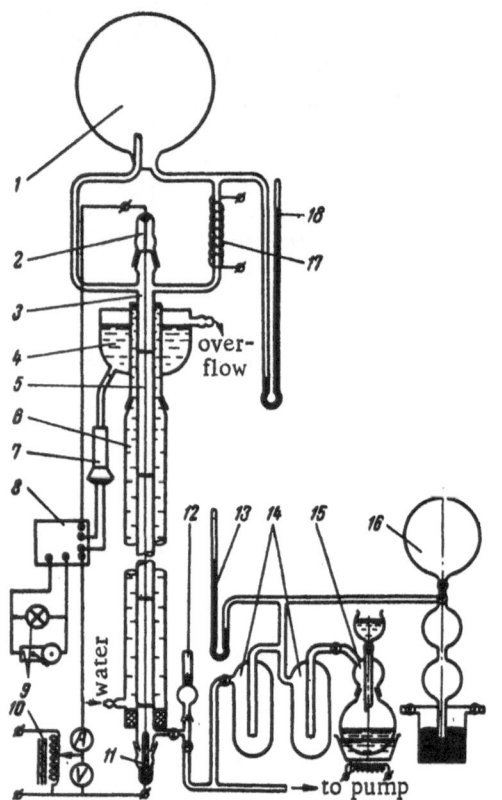

Fig. 2. Plan of low-temperature fractional distillation column: F — imput; P — take-off; W — to waste; A_1 — liquid nitrogen filler; A_2 — nitrogen gas outlet; A_3 — outlet for impurities from condenser; 1) condenser; 2) column with packing; 3) vacuum jacket; 4) shield; 5) still with boiler; 6) oil diffusion pumps; 7) fore-vacuum tank; 8) mechanical pump; 9) ionization pressure gauge; 10) thermocouple vacuum gauge. The operating conditions of the column were controlled by the pressure through the leads F, P, and W.

Fig. 3. Plan of apparatus for the separation of isotopes by the thermal diffusion of CO: 1) reservoir; 2) and 11) electrical leads; 3) column; 4) reservoir for water overflow from condenser; 5) heated filament; 6) condenser; 7) water-operated switch; 8) thyratron relay; 9) signal system (bell and electric lamp); 10) autotransformer; 12) sample ampule; 13) and 18) manometers; 14) traps; 15) apparatus for preparing carbon monoxide; 16) Toepler pump; 17) thermosyphon.

The distillation was normally performed at the normal boiling point of liquid nitrogen. With a column 3.75 m high and 17 mm in diameter, after two days we attained an enrichment in the isotope C^{13} by a factor of 3.65 and in the isotope O^{18} by a factor of 2.46, which corresponds to a theoretical plate height of 2.7 cm. In some experiments, carbon monoxide with $C^{14}O$ was used. An enrichment factor of 13.3 for the isotope C^{14} was obtained and this corresponded to a separation coefficient of $\alpha = 1.02$. The value of the separation coefficient obtained agrees well with theory. The high efficiency of C^{14} enrichment by fractional distillation may make it possible to use this method for increasing the accuracy of determining the age of substances of an organic origin by the carbon-dating method.

To evaluate the capacity of the method and the rate at which columns established a stationary operating state, the transfer of the rare isotope along the column was determined (see Table 5). The retention of liquid was 20% of the total volume of the column. From the relation of the fall in pressure in the column and the heater power it was established that heat losses through the walls of the column were insignificant. The nitrogen consumption for a small metal column was ~30 liters per day. The rate of accumulation of C^{13} in the still of a

TABLE 5

Comparative Efficiencies of Carbon Isotope Separation by Various Methods

Method	Separation coefficient α	HETP, cm	Transfer of rare isotope τ, $g/cm^2 \cdot min$	W, arbitrary units*
Bicarbonate method (1 atm pressure)	1.012	52	$1.7 \cdot 10^{-6}$	1.0
Bicarbonate method (4 atm pressure)	1.0105	15.7	$1.6 \cdot 10^{-6}$	2.7
Fractional distillation of methane on a glass column	1.011	5	$1.8 \cdot 10^{-4}$	8.7
Fractional distillation of carbon monoxide on a glass column	1.011	4	$2.6 \cdot 10^{-4}$	11.0
Fractional distillation of carbon monoxide on a metal column	1.011	2.7	$1 \cdot 10^{-4}$	14.7
Cyanide method**	1.026	10	$1.8 \cdot 10^{-5}$	24.4

* As a measure of efficiency we used the value $W = (\ln F \cdot \tau / L)$, where τ is the transfer of isotope along the column; W is the efficiency.

** Using data from [1].

12-meter column per day with a boiler power of 30 watts was approximately 1.5 C_0 (C_0 is the concentration of C^{13} in the upper reservoir).

Table 5 gives a comparison of the separation efficiencies for carbon isotopes by different methods.

Table 5 shows that fractional distillation of methane or carbon monoxide is considerably more effective than separation by the bicarbonate method. In addition, in the fractionation the carbon in the enriching section is recovered completely by simple evaporation. In the chemical methods (bicarbonate and cyanide) this recovery involves a high consumption of reagents and leads to some noxious waste.

In an ideal cascade, the transfer of the rare isotope should be constant along the cascade. Therefore, to obtain a product with a very high concentration of the rare isotope, the diameter of the column of the last stage should be extremely small, providing that the cascade is not designed for the preparation of large amounts of enriched material. The minimum diameter of the packed column is determined by the dimensions of packing elements and practically cannot be appreciably less than 10 mm. Therefore, if we are considering the preparation of small amounts of product, it is advantageous to use the thermal diffusion method for the last enrichment stages as this is the simplest to construct and to operate. This method was used previously to separate carbon isotopes in methane and required the use of columns consisting of coaxial cylinders. In 1949 we carried out preliminary investigations on the separation of carbon and oxygen isotopes by thermal diffusion of carbon monoxide in columns with hot wires. It was shown that an enrichment in the isotope C^{13} by a factor of more than 3 could be attained on a column 2 m high [13]. In subsequent years we carried out a detailed study of the method. In 1953, a paper was published [14] in which a separation of carbon and oxygen isotopes in a column 5 m high and 12 mm in diameter was described. We investigated the dependence of the enrichment factor and the rate of accumulation of the rare isotope on the wire temperature, column diameter and gas pressure. In some of the experiments, carbon monoxide enriched in the isotopes C^{13} and O^{18} was used. The plan of the thermal diffusion apparatuses we used is illustrated in Fig. 3. The apparatus was made from pyrex glass. It was shown that the enrichment factors for carbon F_C and oxygen F_O isotopes increased with an increase in temperature. Thus, for a column 14.5 mm in diameter and 100 cm high, operating at 428 mm pressure, when the temperature was raised from 500 to 800°, F_C increased from 1.14 to 1.26 and F_O from 1.23 to 1.44. However, raising the temperature above 800° was not advantageous due to carbonization of the platinum filament and, in connection with

125

this, possible breakages in it. The pressure giving the optimal enrichment factor for a column 14 mm in diameter was close to 350 mm and for a column 9 mm in diameter, close to atmospheric. In the latter case the separation coefficient of a column 2 m high after 4-6 hours was close to the maximum value of approximately 3.13. The value of the coefficient H in the equation for transfer of the rare isotope

$$\tau = HN(1-N) - (K_c - K_d)\frac{dN}{dz}$$

in this column was determined from the rate of accumulation of the rare isotope of carbon at the bottom of the column in the initial period, and was equal to $1.9 \cdot 10^{-6} p^4$ g/sec (p is the pressure in atmospheres). The value of H found experimentally agreed well with the calculated value. The value of the parasitic mixing coefficient K_p was determined in the usual way from the relation between the enrichment factor and the pressure [15]. It was found that K_p increased with an increase in the column diameter. Comparison of our data with the results of separating carbon isotopes in the form of methane showed that it was possible to obtain any given enrichment with a shorter column in the case of carbon monoxide, but that a slightly larger consumption of electrical energy was required.

It is interesting to note that while the theoretical value of the ratio $\ln F_C / \ln F_O = 2$, in our experiments it varied between 1.2 and 1.3 when water vapor was absent from the gas. Changes in the gas pressure and the filament temperature did not affect this ratio. The presence of water vapor raised the ratio to 1.5-1.6.

P. S. Petrov, Iu. P. Lapin, A. I. Kasperovich, L. P.Lipikhin, B. Z. Torlin, A. I. Krylov, N. P. Korotkov and V. M. Sadkova participated in separate stages of the work, as well as the authors.

LITERATURE CITED

[1] A. Hutchison, D. W. Stewart and H. C. Urey, J. Chem. Phys. 8, 532 (1940).

[2] A. F. Reid and H. C. Urey, J. Chem. Phys. 11, 403 (1943).

[3] G. G. Deviatykh, Dissertation (Moscow, 1954).

[4] K. Fourhold, J. Phys. Chem. 21, 400 (1924).

[5] G. A. Mills and H. C. Urey, J. Chem. Phys. 8, 403 (1940).

[6] G. G. Deviatykh and A. D. Zorin, J. Phys. Chem. (USSR) 30, 1134 (1956).

[7] N. M. Zhavoronkov and G. A. Iagodin, Doklady Akad. Nauk SSSR 111, 12, 384 (1956).

[8] T. F. Johns, H. Kronberger and H. London, Mass Spectrometry (London, 1952) p. 141.

[9] W. Groth, H. Ihle and A. Murrenholf, Z. Naturforsch. 9A, 805 (1954).

[10] K. Clusius and H. H. Büller, Z. Naturforsch. 9A, 775 (1954).

[11] G. T. Armstrong, F. G. Brickweddle and R. B. Scott, J. Chem. Phys. 21, 1293 (1953).

[12] I. N. Kirschenbaum, Heavy Water [Russian translation] (IL, 1953).

[13] N. N. Tunitskii, G. G. Deviatykh, N. S. Petrov and B. Z. Torlin, J. Theo. Phys. (in press).

[14] H. Sacata, K. Matsuda and E. Tokeda, J. Phys. Soc. Japan 8, 313 (1953).

[15] K. Johns and V. Ferri, The Separation of Isotopes by the Thermal Diffusion Method [Russian translation] (1947).

LOW-TEMPERATURE METHODS FOR SEPARATING HELIUM ISOTOPES (He³ − He⁴)

V. P. Peshkov and V. M. Kuznetsov

In recent years the rare helium isotope He³ has acquired a great deal of value in low-temperature physics. As is well known, in natural helium this isotope is found in percentages less than $10^{-4}\%$ (atmospheric helium) and $10^{-6}\%$ (underground sources); so that the accumulation of He³ in amounts sufficient for experimental purposes is possible only through the development of methods of artificially producing He³ in atomic piles. In order to carry out experiments with He³ − He⁴ mixtures, it is very important to develop methods for separating these isotopes. In the present paper we consider low-temperature methods for separating helium isotopes; the application of these methods provides a rather rapid and reliable method for extracting He³ from a mixture.

Diagram for an He³ − He⁴ Solution

At the present time, the diagram of the liquid-vapor state for He³ − He⁴ solutions is known over a wide range of temperature and pressure.

The most complete data on equilibrium in such a system is that reported by Esel'son and Berezniak [1]. Their results cover the range of concentrations x_3 = 0.4-90.8%, y_3 = 1.9-82.4% (x_3 is the relative molar concentration of He³ in the liquid in terms of the fraction He³/(He³ + He⁴); y_3 is the same quantity referred to the vapor) and are found to be in good agreement with the data reported by Sommers [2].

Measurements in the regions of high concentrations of He³ have been carried out by Peshkov and Kachinskii [3]. The results obtained by these authors are in good agreement with those obtained in [1].

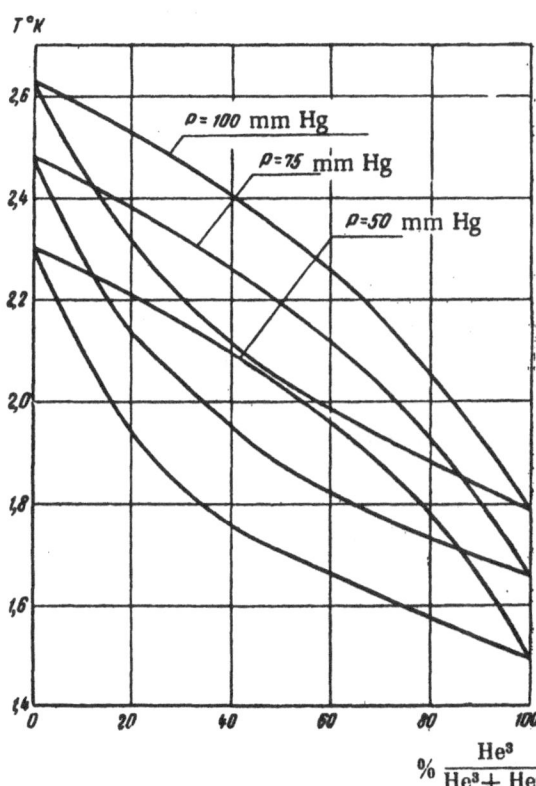

Fig. 1. Isobars for equilibrium between liquid and vapor according to the data of [1, 3].

The two-phase vapor−binary solution of He³ − He⁴ is not ideal; the solution does not obey the general Raoult relation. Deviation from this relation becomes larger at lower temperatures [1]; moreover, the vapor does not satisfy the equation of state for an ideal gas [4]. A calculation of the second virial coefficients shows that the departure from an ideal gas increases as the density is increased. At temperatures below approximately 0.8°K a separation of the liquid into two phases is observed [5]. Since there are no isentropic points for the region of higher temperatures of the solution (Fig. 1), the equilibrium concentration of He³ in the vapor above a two-phase solution is always higher than in the liquid. The solubility curve for He³ in He⁴ [5], and the lines for the λ-transitions [6-8] are shown in Fig. 2 (the meaning of the dashed curves is explained below).

127

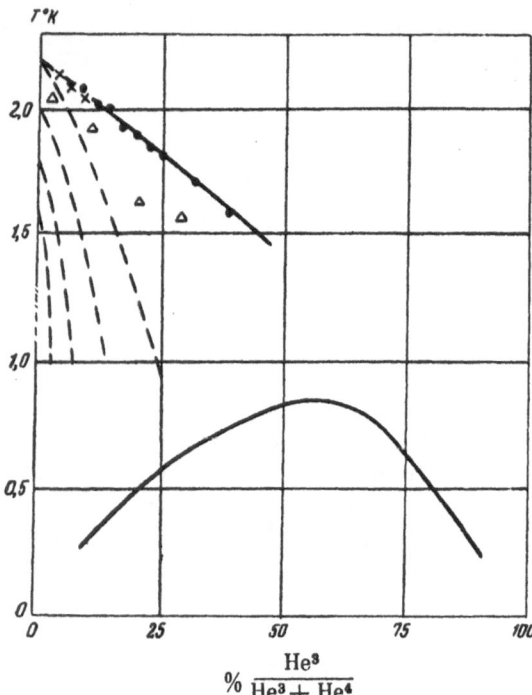

Fig. 2. Curves for solubility of He[3] in He[4] [5], λ -transition lines in He[3] − He[4] solutions (Δ − [6]; ● − [7]; × − [8]) and equilibrium concentrations of the liquid for enrichment of the solution by thermal osmosis (dash lines).

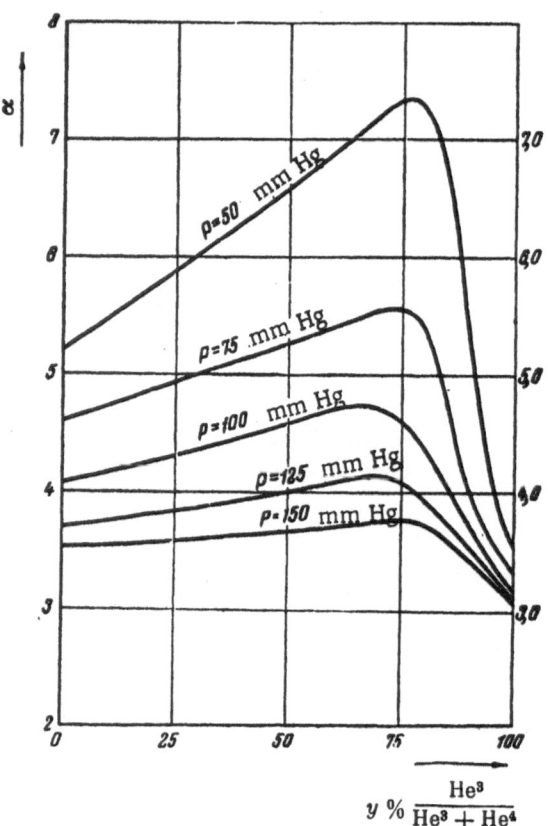

Fig. 3. The separation coefficient α as a function of vapor concentration at various pressures

The experimental data [1,3] allow us to compute the separation coefficient

$$\alpha = \frac{y_3(1-x_3)}{x_3(1-y_3)}. \tag{1}$$

The curve for α as a function of y_3 for different pressures has a characteristic maximum in the concentration region of 70-80% (Fig. 3). With $1-x_3 = x \le 0.5$, the following relation can be used

$$x = \bar{\alpha}y = \bar{\alpha}(1-y_3), \tag{2}$$

where $\bar{\alpha} = 3.1$. With accuracies up to $\pm 20\%$, $\bar{\alpha}$ can be considered constant in the temperature region 1.5-2°K.

Enrichment of Helium in the He[3] Isotope by Means of the Mechano-Caloric Effect ("Thermal Osmosis")

The superfluidity of He-II makes it possible to enrich a mixture using thermal flow [9-11] by flowing He[4] in a film [12,13], or through a fine filter [9, 14,15]. It has been shown experimentally that the best results are obtained by "thermal osmosis," which provides reliable extraction of He[3] and rather good reproducibility [9,14].

The enrichment coefficient for superfluid filtration can be estimated [9] on the basis of the following considerations. If it is assumed that the superfluid flow and diffusion occur across the same effective cross section of the filter, and that the velocity of the superfluid flow is limited to the critical velocity of 20 cm/sec, the superfluid flow through unit area is

$$\omega_1 = \rho_s v_s,$$

while the He[3] flow, determined by diffusion is

$$\omega_2 = D\rho \frac{m_3 x_3^0}{m_4 l},$$

where D is the diffusion coefficient for He^3; l is the length of the filter; ρ_s is the density of the superfluid part of the helium; x_3^0 is the molar concentration before the filter; and m$_3$ and m$_4$ are the masses of He^3 and He^4 atoms, respectively.

The concentration of the transmitted helium

$$x_3 = \frac{\omega_2 m_4}{\omega_1 m_3} = \frac{D\rho x_3^0}{l\rho_s v_s},$$

i.e., the limiting value of the enrichment coefficient, is

$$A = \frac{x_3^0}{x_3} = \frac{l\rho_s v_s}{D\rho}. \tag{3}$$

The diffusion coefficient for He^3 in He-II has been experimentally determined by Beenakker et al. [16]. It is found to fall rapidly from 10^{-2} cm^2/sec at 1.3°K to 10^{-3} cm^2/sec at 1.6°K, and $4 \cdot 10^{-5}$ cm^2/sec at 2.1°K. Thus, with a filter length l = 5 cm, the separation coefficient at 1.3°K is A = 10^4; at 1.6°K, A = $8 \cdot 10^4$; and at 2°K, A = $6 \cdot 10^5$; while at 2.1°K, A = $5 \cdot 10^5$, i.e., the enrichment coefficient is a maximum at approximately 2°K. In practice, because of discontinuities in extended diffusion, due to lack of superfluid, and because of imperfections in the filter, the enrichment is always smaller.

The limiting enrichment for the filter can be estimated as follows [9]; if the He^3 concentration is not very high, the following relation given by Pomeranchuk [17] applies for thermal osmosis:

$$-SdT + \frac{dp}{\rho} - \frac{k}{m_4} d(x_3 T) = 0,$$

where S is the entropy per gram of He^4; ρ is the density; p is the pressure; m$_4$ is the mass of the He^4 atom; and k is the Boltzmann constant.

In the temperature region above 1°K the entropy of He-II can be expressed by the formula

$$S = 0.405 \left(\frac{T}{2.19}\right)^{5.5} \quad \text{cal/g} \cdot \text{deg},$$

i.e.,

$$SdT = \frac{1}{6.5} d(ST).$$

The density of He-II does not vary greatly so that the expression can be written in the form

$$d\left\{\left(\frac{k x_3}{m_4} + \frac{S}{6,5}\right) T - \frac{p}{\rho}\right\} = 0$$

or, along the thermal-osmosis path,

$$\left(\frac{k x_3}{m_4} + \frac{S}{6,5}\right) T - \frac{p}{\rho} = \text{const}.$$

Since in the extraction of He^3 by thermal osmosis the concentration at one side is essentially zero, then

$$\left(\frac{k x_3}{m_4} + \frac{S}{6,5}\right) T - \frac{p}{\rho} = \frac{S_0 T_0}{6,5} - \frac{p_0}{\rho}$$

or

$$x = \frac{m_4}{kT} \left(\frac{S_0 T_0 - ST}{6,5} - \frac{p_0 - p}{\rho}\right). \tag{4}$$

The values of x_3 computed by the above formula for various values of T are plotted on the curve (Fig. 2) in dashed lines. As is apparent from the curve, thermal osmosis does not go to conveniently large concentrations so that the enrichment of mixtures with concentrations of He^3 greater than 10%, and the removal from He^4 are most conveniently carried out by rectification.

Rectification

At small velocities of the phases an estimate of the parameters of a rectification column can be carried out if it is assumed, as has been done by Westhaver [18] and Kuhn [19], that the motions of the liquid phase and the gas phase are laminar. For a column consisting of a tube of radius $\underline{a}$ along the walls of which flows a liquid film of thickness δ, where $\delta \ll a$, for stationary conditions we may write for the vapor phase

$$\frac{D_y}{r} \frac{\partial}{\partial r}\left(r \frac{\partial y}{\partial r}\right) + D_y \frac{\partial^2 y}{\partial z^2} - v \frac{\partial y}{\partial z} = 0 ; \tag{5}$$

and for the liquid phase

$$D_x \left(\frac{\partial^2 x}{\partial r^2} + \frac{\partial^2 x}{\partial z^2}\right) - u \frac{\partial x}{\partial z} = 0, \tag{6}$$

where $\underline{r}$ is the radius; $\underline{z}$ is a coordinate taken along the axis of the tube in the downward direction; $\underline{y}$ and $\underline{x}$ are the relative molar concentrations of He^4; D_y and D_x are the coefficients of molecular diffusion in the vapor and liquid phases, respectively; $\underline{v}$ is the linear velocity of the vapor; and $\underline{u}$ is the linear velocity of the liquid. We denote the average concentrations in the flow by $\overline{y}$ and $\overline{x}$.

$$\overline{y} = \frac{2}{a^2 v_0} \int_0^a \left(vy - D_y \frac{\partial y}{\partial z}\right) r dr, \tag{7}$$

$$\overline{x} = \frac{1}{\delta u_0} \int_0^\delta \left(ux - D_x \frac{\partial x}{\partial z}\right) dr , \tag{8}$$

and the molar densities of the vapor and liquid phases by n_y, and n_x; the mean velocities by v_0 and u_0; and the vapor taken to the upper part of the column by γ ; then, from mass balance considerations,

$$2\delta u_0 n_x - a v_0 (1 - \gamma) n_y = 0 \tag{9}$$

and

$$\overline{y} - \overline{x}(1 - \gamma) = \overline{y}_0 \gamma = \overline{x}_0 \gamma, \tag{10}$$

where $\overline{y}_0$ is the mean concentration of the vapor which is removed (the molar flows of the liquid and vapor are assumed constant over the height of the column).

Assuming that the distribution of velocities and D_y / v_0 is independent of $\underline{z}$, From Eqs. (5), (6), (7) and (8) we have

$$\frac{\partial \overline{y}}{\partial z} = \frac{2}{a^2 v_0} \int_0^a \left(v \frac{\partial y}{\partial z} - D_y \frac{\partial^2 y}{\partial z^2}\right) r dr = \frac{2 D_y}{a v_0}\left(\frac{\partial y}{\partial r}\right)_i$$

$$\frac{d \overline{x}}{\partial z} = \frac{D_x}{\delta u_0}\left(\frac{\partial x}{\partial r}\right)_i;$$

where the subscript "i" means that the derivative is taken over the boundary between the vapor and liquid phases. Assuming that in the first approximation

$$\left.\begin{aligned}
\left(\frac{\partial y}{\partial r}\right)_i &= 2k_y \frac{y_i - \overline{y}}{a} , \\
\left(\frac{\partial x}{\partial r}\right)_i &= 2k_x \frac{x_i - \overline{x}}{\delta} ,
\end{aligned}\right\} \tag{11}$$

we have

$$\frac{a^2 v_0}{4 D_y k_y} \frac{d\bar{y}}{dz} = y_i - \bar{y} ; \tag{12}$$

$$\frac{\delta^2 u_0}{2 D_x k_x} \frac{d\bar{x}}{dz} = x_i - \bar{x} , \tag{13}$$

where k_x and k_y are factors of order unity.

Using Eqs. (2), (10), (12), (13), and (9), we can write

$$\bar{x} = \frac{(\bar{\alpha}-1) e^{\beta z} - \bar{\alpha}\gamma}{\bar{\alpha} - 1 - \bar{\alpha}\gamma} \bar{x}_0, \tag{14}$$

where

$$\frac{1}{\beta} = - \frac{\bar{\alpha}(1-\gamma) v_0 a}{4 [\bar{\alpha}(1-\gamma)-1]} \left(\frac{\bar{\alpha}}{D_y k_y} + \frac{\delta n_y}{\bar{\alpha} D_x k_x n_x} \right). \tag{15}$$

The quantity β is positive since v_0 is negative. As is apparent from the expression for β, removal of γ leads to an apparent reduction in the separation coefficient $\alpha' = \bar{\alpha}(1-\gamma)$. Eqs. (14) and (15) apply only in the region of concentrations in which $x = \bar{\alpha}y$.

In the nonselective scheme for the case in which the resistance to mass transfer in the liquid can be neglected, i.e., $\bar{x} = x_i = x$, using the general relation in (1), we have [20]:

$$\frac{1}{\beta} = - \frac{\alpha a^2 v_0}{(\alpha - 1)4 D_y k_y} , \tag{16}$$

$$\frac{\bar{x}}{(1-\bar{x})^{1/\alpha}} = \frac{\bar{x}_0}{(1-\bar{x}_0)^{1/\alpha}} e^{\beta z}. \tag{17}$$

Eq. (17) applies for any concentration if (1) is valid with constant α. By analogy with heat transfer, for turbulent motion of the vapor we have

$$Nu' = 0.023\, Re^{0.8} (Pr')^{0.4} . \tag{18}*$$

Here

$$Nu' = \frac{k_g d}{D_y} \tag{19}$$

is the Nusselt number where $k_g = j/n_y \Delta y$ is the coefficient of mass transfer in the vapor [22] (cm/sec); j is the diffusion flow at the phase boundary in mol/cm$^2 \cdot$ sec; Δy is taken as $y_i - \bar{y}$; and Pr' is the Prandtl number. At the phase boundary the following relation obtains:

$$\frac{j}{n_y} = k_r \Delta y = D_{eff} \left(\frac{\partial y}{\partial r} \right)_i , \tag{20}$$

where $D_{eff} = D_y + D_{tur}$ (D_{tur} is the coefficient for turbulent diffusion in cm^2/sec).

Then, from Eqs. (11), (19), and (20) we have

$$D_{eff} = Nu' D_y \frac{1}{4 k_y} .$$

Substituting in Eq. (16) D_{eff} in place of D_y, we find $1/\beta$ for the turbulent mode

$$\frac{1}{\beta} = \frac{a}{0.046} \frac{\alpha}{\alpha - 1} Re^{0.2} (Pr')^{0.6}. \tag{21}$$

---

*In the case of heat transfer this equation is of the form [21]:

$$Nu = 0.023\, Re^{0.8} Pr^{0.4},$$

where $Nu = \kappa d/\lambda$, (κ is the heat transfer coefficient in cal/cm$^2 \cdot$ sec $\cdot$ °C, λ is the thermal conductivity in cal/cm $\cdot$ sec $\cdot$ °C), $Pr = \nu/a$, where $\underline{a}$ is the coefficient of temperature conductivity in cm^2/sec.

Applied to a nozzle column $Re = \dfrac{|v_0| d_{equiv}}{\psi \nu}$ where $d_{equiv} = \dfrac{4\psi}{f'}$ is the equivalent diameter [23]; ψ is the free volume of the nozzle (that fraction of the volume of the column which is not occupied by the material in the nozzle); f' is the surface of the nozzle per 1 cm^3 of the column volume. Setting $a = d_{equiv}/2$, from Eq. (21) we have

$$\frac{1}{\beta} = \frac{d_{equiv}}{0,092} \frac{\alpha}{\alpha - 1} Re^{0,2} (Pr')^{0,6}. \tag{22}$$

If, in place of Eq. (18) we use the criteria equations for mass transfer in a granular layer [24,25], an equation of the same form but with different constants appears. The quantity $1/\beta$ can also be expressed in terms of the height of a transfer unit (HTU).

By definition [26],

$$(HTU)_g = \frac{l}{\displaystyle\int_{y_l}^{y_0} \frac{dy}{y_i - y}},$$

where l is the height of the section of column at the ends of which the vapor concentration is y_0 and y_l; $y_i - y$ is the diffusion pressure in the radial direction; y is the concentration of vapor in the core of the flow (in the turbulent mode $y \approx \bar{y}$).

Neglecting the resistance to mass transfer in the liquid, for a film column, taking account of (18) we find [27]:

$$(HTU)_g = \frac{a}{2 \cdot 0,023} Re^{0,2} (Pr')^{0,6}.$$

Comparing this with (20) we find

$$\frac{1}{\beta} = \frac{\alpha}{\alpha - 1} (HTU)_g. \tag{23}$$

The same relation is obtained for a nozzle column. The distribution of concentrations (14) and (17) makes it possible to determine the He^4 content in the column (as before, the resistance to mass transfer in the liquid is neglected):

$$q = \frac{\sigma}{l} \cdot \int_0^l \bar{x} dz ; \tag{24}$$

here q is the number of moles of He^4 in the column, and σ is the number of moles of the mixture in the column.

When working with extraction in the concentration region $x \leq 0.5$ ($x = \bar{\alpha} y$) for small values of x_0/x_l (x_l is the He^4 concentration in the lower part of the column) from Eq. (24) we have [20]:

$$q = \frac{\sigma \bar{x}_l}{\beta l}. \tag{25}$$

Under these conditions, in the case of periodic rectification the mean concentration of He^4 in the output

$$\bar{x}_{av} = \frac{1}{\Sigma_0} \int_{\Sigma - \Sigma_0}^{\Sigma} \bar{x}_0 d\Sigma ,$$

where Σ is the loading of the unit in moles, Σ_0 is the amount of the separated product, can be computed from the formula [20]:

$$\bar{x}_{av} = \frac{\left(1 - \dfrac{\bar{\alpha}\gamma}{\alpha - 1}\right) Q e^{-\beta' l}}{\bar{\alpha} \Sigma_0} \ln \frac{\bar{x}_2}{\bar{x}_1} , \tag{26}$$

where Q is the amount of He^4 in the unit; $\bar{x}_1$ and $\bar{x}_2$ is the concentration of He^4 in the unit at the beginning and at the end of removal.

In Fig. 4 is shown a system for enriching natural helium in the He^3 isotope. The enrichment in the light isotope takes places as a result of the simultaneous effects of thermal osmosis and rectification.

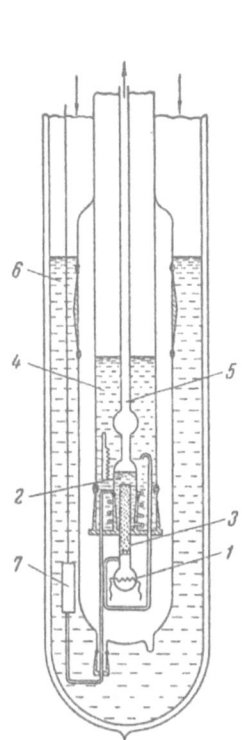

Fig. 4. Apparatus for enrichment of natural helium in the He^3 isotope.

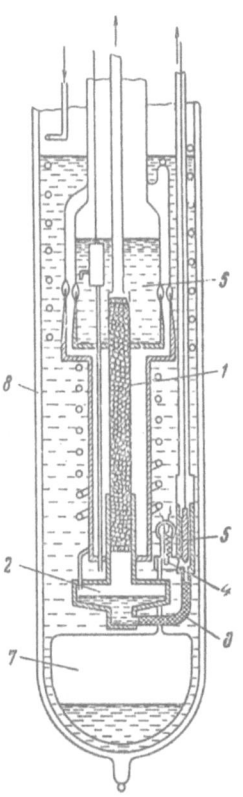

Fig. 5. Apparatus for concentration of mixtures and production of He^3: 1) nozzle column in vacuum container; 2) vat; 3) filter; 4) heater; 5) valves controlling filter output; 6) helium bath for cooling upper part of column; 7) storage tank for helium which flows through filter; 8) outer helium bath.

Because of the heat which is generated by the heater 1, helium from container 2 undergoes superfluid flow through filter 3 into the inner Dewar 4. Further enrichment takes place as a result of rectification in tube 5. The storage container 2 is filled with helium from the outer Dewar 6 through valve 7.

The system can be used to prepare up to 3 nm^3 of helium* per hour and is designed for continuous operation, in contrast with the apparatus used by Esel'son and Lazarev [14] which operates cyclically. In practice the enrichment coefficient is approximately $2 \cdot 10^4$, which is close to the calculated value as computed with Eq.(3).

A determination of the capacity of a filter for enriching solutions with concentrations of the order of several percent by thermal osmosis indicates [28] that for any capacity a higher enrichment corresponds to a higher temperature drop if one end of the filter is kept at a fixed temperature, i.e., a relation of the form given in (4) is observed qualitatively.

The degree of extraction of He^3 through a filter 45 mm long is approximately 99.98%.

A device which operates on the same principle as the above system and which is used for concentration of mixtures and to obtain He^3 is shown in Fig. 5. In this system a column 20 cm long and 1 cm in diameter is filled with a nozzle of wire rings 2 mm in diameter (the wire is 0.15 mm in diameter).

---

*At normal pressure and temperature, signified here by the letter n— Publishers note.

The processing of a mixture is carried out in two cycles. In the first cycle the mixture is extracted at the most convenient conditions of operation of the column and then the He^4 is extracted from the vat residue by thermal osmosis ($T_{V\,at} \simeq 2°K$, the capacity of the filter is 100 nl/hr). The mean concentration of the semi-finished product is 30-50%.

In the second cycle He^3 is obtained by rectification of the semi-finished product. The removal rate in the second cycle varies from 10 (at the beginning of removal) to 4 (end of removal) nl/hr.

In this way 4 nl of He^3 with a purity of 99.995% is obtained; in this case, at the end of the cycle the column contains approximately 0.25-0.3 nl of He^3 and 0.8 nl of He^4. The vapor velocity in the column is limited by the quantity v_{cr}, i.e., the velocity at which wetting occurs. As is well known, for a given vapor velocity the pressure drop along a wet column is much higher than along a dry column.

As has been shown by P. L. Kapitza [29], under certain given conditions the wave motion of a thin layer of viscous fluid is more stable than laminar motion. Under these conditions, taking account of the breakup of the gas flow along the wave surface of the liquid yields the possibility of not only quantitatively determining the increase in the pressure drop in a wet tube, but also the possibility of computing the critical vapor velocity with rather high accuracy. An experimental determination of v_{cr} for a nozzle column [28] yields the same dependence as that obtained theoretically [29].

A computation of the efficiency of the column using Eq. (26) (as has been shown in [20] the resistance to mass transfer in the liquid can be neglected) yields $e^{\beta'l} \simeq 4000$, i.e., $1/\beta \approx 2.5$ cm. The correct order of magnitude of this quantity is also obtained using Eq. (22) (since γ is small, $\beta' \approx \beta$).

The time required to establish these conditions in the column is approximately 10 seconds, i.e., close to the calculated value [20]:

$$\tau_{calc} = \frac{4a\sigma}{(\alpha - 1)\,w} \,,$$

where $\underline{w}$ is the flow of vapor or liquid in moles/sec.

SUMMARY

Low-temperature methods of separating He^3 — He^4 mixtures are more effective, convenient, and reliable than methods which involve separation of these helium isotopes at ordinary temperatures.

LITERATURE CITED

[1] B. N. Esel'son and N. G. Berezniak, J. Exptl.-Theoret. Phys. (USSR) 30, 628 (1956).

[2] H. S. Sommers, Phys. Rev. 88, 113 (1952).

[3] V. P. Peshkov and V. N. Kachinskii, J. Exptl.-Theoret. Phys. (USSR) 31, 720 (1956).

[4] J. E. Kilpatrick, W. E. Keller and E. F. Hammel, Phys. Rev. 94, 1103 (1954).

[5] G. K. Walters and W. M. Fairbank, Phys. Rev. 103, 262 (1956).

[6] B. M. Abraham, B. Weinstock and D. W. Osborne, Phys. Rev. 76, 864 (1949).

[7] B. N. Esel'son, N. G. Berezniak and M. I. Kaganov, Doklady Akad. Nauk SSSR 111, 568 (1956).

[8] J. G. Dash and R. D. Taylor, Phys. Rev. 99, 598 (1955).

[9] V. P. Peshkov, J. Exptl.-Theoret. Phys. (USSR) 30, 850 (1956).

[10] C. T. Lane, H. A. Fairbank, L. T. Aldrich and A. O. Nier, Phys. Rev. 73, 256 (1948).

[11] C. A. Reynolds, H. A. Fairbank, C. T. Lane, B. B. McInteer and A. O. Nier, Phys. Rev. 76, 64 (1949).

[12] J. Daunt, R. Probst, H. Johnston, L. Aldrich and A. O. Nier, Phys. Rev. 72, 502 (1947).

[13] B. N. Esel'son, B. G. Lazarev and I. M. Lifshits, J. Exptl. Theoret. Phys. (USSR) 20, 749 (1950).

[14] B. N. Esel'son and B. G. Lazarev, J. Exptl.-Theoret. Phys. (USSR) 20, 743 (1950).

[15] K. Atkins and D. R. Lovejoy, Can. J. Phys. 32, 702 (1954).

[16] J. J. Beenakker, K. W. Takonis, E. A. Lynton, Z. Dokoupil and G. van Soest, Physica 18, 433 (1952).

[17] I. Pomeranchuk, J. Exptl.-Theoret. Phys. (USSR) 19, 42 (1949).

[18] J. M. Westhaver, Ind. Eng. Chem. 34, 126 (1942).

[19] W. Kuhn, Helv. Chim. Acta 25, 252 (1942).

[20] V. P. Peshkov, J. Tech. Phys. (USSR) 26, 664 (1956).

[21] M. A. Mikheev, Fundamentals of Heat Transfer (State Power Press, 1949) p. 98.

[22] D. A. Frank-Kamenetskii, Diffusion and Heat Transfer in Chemical Kinetics (AN SSSR, 1947) pp. 24, 171.

[23] V. M. Ramm, Absorption Processes in the Chemical Industry (State Chemical Press, 1951) p. 188.

[24] M. E. Aerov and N. I. Umnik, J. Tech. Phys. (USSR) 26, 1223 (1956).

[25] C. N. Satterfield and H. Resnick, Chem. Eng. Progr. 50, 504 (1954).

[26] A. P. Colburn, Trans. Am. Inst. Chem. Eng. 35, 211 (1939).

[27] I. P.Usiukin and L. S. Aksel'rod, Oxygen 3, 1 (1952).

[28] V. M. Kuznetsov, J. Exptl.-Theoret. Phys. (USSR) 32, 1001 (1957).

[29] P. L. Kapitza, J. Exptl.-Theoret. Phys. (USSR) 18, 3, 19 (1948).